189
Advances in Anatomy Embryology and Cell Biology

Gundela Meyer

Genetic Control of Neuronal Migrations in Human Cortical Development

With 29 Figures

 Springer

Gundela Meyer
Departamento de Anatomía Faculdad de Medicina
Universidad de La Laguna
38071 La Laguna/Tenerife
Canary Islands
Spain
e-mail: gmeyer@ull.es

ISSN 0301-5556
ISBN-10 3-540-36688-1 Springer Berlin Heidelberg New York
ISBN-13 978-3-540-36688-1 Springer Berlin Heidelberg New York

Springer is a part of Springer Science+Business Media
springer.com
© Springer-Verlag Berlin Heidelberg 2007

Editor: Simon Rallison, Heidelberg
Desk editor: Anne Clauss, Heidelberg
Production editor: Nadja Kroke, Leipzig
Cover design: WMX Design Heidelberg
Typesetting: LE-TEX Jelonek, Schmidt & Vöckler GbR, Leipzig
Printed on acid-free paper   SPIN 11802631   27/3150/YL – 5 4 3 2 1 0

List of Contents

Acknowledgements

I want to thank Alfredo Cabrera-Socorro and Carolina Acosta Hernandez for helping with the photographs and the bibliography of this monograph. I also gratefully acknowledge the editorial help of Prof. Francisco Clascá.

Abbreviations

AChE	Acetylcholinesterase
ApoER2	Alipoprotein E receptor type 2
CR cell	Cajal–Retzius cell
CS	Carnegie stage
Dab1	Disabled 1
DCX	Doublecortin
E	Embryonic day
GW	Gestational week
IZ	Intermediate zone
MZ	Marginal zone
SGL	Subpial granular layer
SVZ	Subventricular zone
V	Ventricular zone
VLDLR	Very low density lipoprotein receptor

Abstract

The early steps in corticogenesis are decisive for the correct unfolding of neurogenesis, neuronal migration and differentiation under tight genetic control. In this monograph, we outline the main events in human preplate formation, the gradual transformation of the preplate into the cortical plate, and the establishment of the transient compartments of the foetal cortical wall. The main neuronal populations of the embryonic and fetal cortex are presented according to their timetable of appearance and the expression of developmentally relevant gene products, with the main focus on members of the Reelin-Dab1 signalling pathway, LIS1 and Doublecortin, all of which are crucial for cortical migration. The often significant developmental differences between the lissencephalic rodent brain, which has become the prevailing model of corticogenesis, and the highly differentiated, gyrated human brain are pointed out and discussed.

1
Introduction

Cortical development is a complex, tightly regulated process that eventually leads to the six-layered adult human neocortex, the substrate of the unique cognitive, emotional, and social abilities of our species. The basic mechanisms of early cortical development are believed to be very similar among mammals, which has led to a tendency of extrapolating experimental data from rodents on humans. Certain traits, such as the general pattern of forebrain regionalization and expression of region-specific genes, seem to be conserved among vertebrates (e.g., Kammermeier and Reichert 2001; Puelles et al. 2000; Smith Fernandez et al. 1998; Abu-Kahlil et al. 2004). Furthermore, the idea of a basic uniformity of neocortical structure in all mammals, proposed by Rockel et al. (1980), has found wide acceptance. One of the recurrent themes of this monograph deals with exactly these questions: How similar are the developmental processes in lower mammals and in primates, in particular, in humans? Are there peculiarities in human cortical development which do not exist, or are not easily recognizable, in rodents?

One of the most noticeable changes during evolution has been the dramatic increase in brain size, linked to the capacity to generate more neurons. The prolonged period of neurogenesis in anthropoid primates and humans makes possible a higher number of mitotic cycles, so that each dividing progenitor cell undergoes more rounds of cell divisions (Kornack and Rakic 1998). In mice, cortical progenitors undergo 11 rounds of cell division (Takahashi et al. 1995), whereas in the macaque monkey there are at least 28 mitotic rounds (Kornack and Rakic 1998), and probably even more in humans. Multiple genes play a critical role in controlling normal brain size, and perturbations of their functions may lead to a condition known as microcephaly (small brain). One form of primary microcephaly is Microcephalia vera, with no other associated malformation but a small brain (50% or less of its normal mass) with a relatively conserved gyral pattern and brain architecture comparable to the brains of early hominids such as *Australopithecus africanus*. Four microcephaly genes (*ASPM*, *microcephalin*, *CDK5RAP2*, and *CENPJ*) have been cloned recently (Bond et al. 2002, 2005; Jackson et al. 2002; Trimborn et al. 2004), and they encode proteins which seem to specifically control neuron number in the developing human brain, acting at different levels of the "mechanics" of mitotic division (reviewed by Francis et al. 2006; Hill and Walsh 2005). The *ASPM* and *microcephalin* genes experienced strong positive selection in the ape lineage leading to the human and may have contributed to the evolutionary enlargement of the human brain (Evans et al. 2004a, b). The elevated expression levels of relevant brain genes has been crucial for human evolution (Cáceres et al. 2003; Uddin et al. 2004), and the analysis of their normal expression pattern in the developing human cortex may provide insights into their possible role in the molecular Bauplan of the brain.

Importantly, the increase in size of the human cortex is accompanied by an increase in morphological complexity, as already recognized by Ramón y Cajal

(1917) and beautifully illustrated in his book (1911), or as described in the extensive Golgi studies of Poljakow (1979) of human cortex development. Our own studies in developing and adult brains of diverse mammalian species, including rodents, carnivores, nonhuman primates, and humans, conveyed the impression of an increasing complexity of cortical structure, reflected by a progressive diversification of discrete cytoarchitectonic areas, an increasing intricacy of intracortical connectivity, and an immense variety of interneuron axonal arborizations in cat and human cortex, compared to the rodent cortex (Meyer 1983, 1987; Meyer and Ferres Torres 1984, Meyer and Wahle 1988; Wahle and Meyer 1987; 1989). The extraordinary structural complexity of the developing human brain is already recognizable at the earliest stages of cortical development (Meyer et al. 2000), and has been described exhaustively in the case of the Cajal–Retzius cells in the marginal zone (MZ; Meyer and Gonzalez Hernandez 1993; Meyer and Goffinet 1998; Meyer et al. 1999; Meyer et al. 2002a). As noted before (Meyer 2001), the cortex develops as a whole, and transient cortical compartments such as the MZ and the subplate, but also the intermediate and subventricular zones, evolve in parallel with the cortical plate and reach maximal differentiation in human. The distinct compartments of the developing cortical wall closely interact and function in concert. The cell composition of the individual compartments will be addressed and related to the expression of the gene products, which we have studied in each compartment.

Of the countless gene products that regulate cortical development (reviewed by Guillemot et al. 2006), we focus in our monograph on those more closely studied by us in previous work on human cortical development, all of which are crucial for the migration of neurons from their place of birth into the cortical plate: Reelin, its receptors and the cytoplasmic adapter protein Disabled 1 (Dab1), which form the Reelin–Dab1 signaling pathway, the microtubule-associated proteins LIS1 and Doublecortin (DCX), and the tumor-suppressor p73. When introducing these gene products, we will compare the human expression pattern with the data available from rodent studies, trying to point out similarities and divergences. The comparative aspect is of uppermost importance in the cases of Reelin, LIS1, and DCX, because in human, mutations of these genes lead to severe cortical malformations that belong to the spectrum of type1 lissencephalies, characterized by a complete loss or simplification of the cortical gyri and sulci, along with a severe disturbance of neuronal migration, whereas the phenotypes of the respective mouse mutants are much less dramatic (reviewed by Francis et al. 2006). The comparison between the expression patterns of LIS1 and DCX in the normal fetal brain and the human cortical phenotype of identified mutations of these genes are presented in the last chapters of this monograph. Further identification of genes and molecular pathways involved in cortical migration, together with a systematic analysis of their expression patterns directly in human samples may lead to new insights into the cellular and molecular mechanisms that control the formation of the gyrated human cortex.

2
Materials and Methods

Our human brain material used in the papers referred to in this monograph was collected over the last 20 years and belongs to the collection of fetal and adult brains of the Department of Anatomy of the University of La Laguna. The embryonic and fetal brains were obtained from spontaneous or medically induced abortions in accordance with the Spanish legislation and supervised by the Ethical Committee of our institution. The adult brains came from autopsies or donations to the Department of Anatomy. The embryonic material, 10 cases, 5–8 gestational weeks (GW) old, was classified according to Carnegie stages (CS; O'Rahilly and Müller 1994). Fetal brains aged from 9 to 40 GW included several cases per gestational week, with a total number of 45 cases. Most brains were fixed in Bouin, others in Carnoy or formaldehyde, embedded in paraffin using standard protocols, and cut into serial 10 μm thick sections, mostly in a coronal plane. Tissue blocks dissected from the frontal or temporal lobe of infants and adults were processed in the same way. The mutant mice and human brain mutations presented in this monograph are described in detail in the original papers.

Immunohistochemistry was carried out using a diversity of primary antibodies, most notably against Reelin (de Bergeyck et al. 1998; ab 142, gift of André Goffinet), Dab1 (ab L2, gift of André Goffinet and ab B3, gift of Brian Howell), VLDLR and ApoER2 (gift of André Goffinet), p73 (gift of Daniel Caput), Tbr1 (gift of Robert Hevner), DCX (gift of Joseph Gleeson). The commercial antibodies against calretinin and calbindin were obtained from Swant (Bellinzona, Switzerland). The LIS1antibody (SC 7577) was from Santa Cruz Biotechnology (CA, USA). The protocols for single and double immunolabeling and in situ hybridization were described in the original papers.

3
The Anatomical Basis of Human Cortical Development

3.1
The Embryonic Period: Formation of the Cortical Anlage and Early Regionalization

We summarize here the main events in telencephalic development during the embryonic period proper, i.e., the first 8 postovulatory weeks, following the detailed timetable of the appearance of outstanding anatomical landmarks established by O'Rahilly and Müller (1994). (For further details, see Kostovic 1990a; Sidman and Rakic 1973, 1982). In human embryology, developmental steps are expressed as Carnegie stages (CS), which are based on external and internal morphological criteria and thus more reliable than gestational ages or body size (O'Rahilly and Müller 2000). One CS covers 2 or 3 days and is defined by the appearance of certain organs and structures.

The neural tube begins to fuse at CS 10 ($\approx$22 days) in the rhombencephalic and spinal regions. The tube remains open at the anterior neuropore, which closes during the 4th week, at CS 11. At this stage, the lamina terminalis and commissural plate become apparent at the floor of the prosencephalon, the most rostral portion of the developing brain. At CS 13, both the rostral and caudal neuropores are closed, and the future ventricular system is no longer in communication with the amniotic cavity. It contains an "ependymal fluid" which is probably formed by the cells lining the ventricles, since the choroid plexuses appear later. At this moment, the telencephalon consists only of a small telencephalon medium or impar that

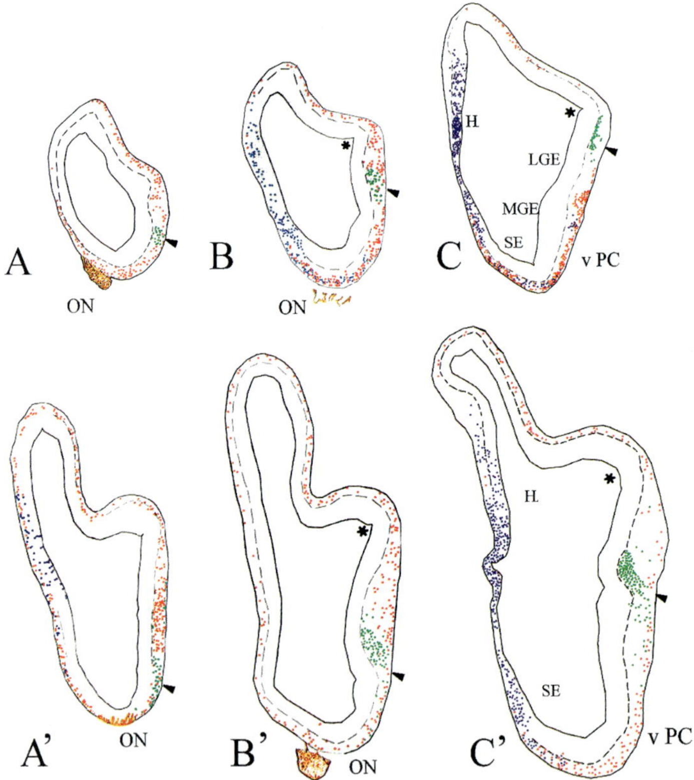

Fig. 1A–F Diagrams of coronal sections at Carnegie stages 17 (**A–F**) and 18 (**A'–F'**), from rostral (**A, A'**) to caudal (**F, F'**). *Dots* represent immunostained cells; *green*: calretinin, *blue*: calbindin, *red*: Reelin. In *brown*: the calretinin-positive fibers from the olfactory placode to the future olfactory bulb. The *asterisks* indicate the pallio-striatal angle, while the *arrowheads* point to a calretinin-positive migratory stream from the lateral ganglionic eminence to the prospective paleo-pallium. *CE*, caudal eminence; *H*, cortical hem; *HT*, hypothalamus; *LE*, lamina epithelialis; *LGE*, lateral ganglionic eminence; *MGE*, medial ganglionic eminence; *SE*, septal eminence; *T*, thalamus; *vPC*, ventral prosencephalon. Note the intense immunopositivity in the cortical hem, as well as the predominance of Reelin immunoreactivity in the ventral part of the hemisphere at this early preplate stage

lies rostral to the anlage of the chiasma. Concurrently, the terminal–vomeronasal crest appears at both sides of the facial mesenchyme, and cells leave the crest to form the future vomeronasal and terminal ganglia. Late in the 5th week, at CS 14, the telencephalic vesicles begin to evaginate bilaterally. They are separated in the midline by the roof plate, which continues rostrally as the lamina terminalis and the commissural plate (lamina reuniens), the site where the future cerebral commissures will begin to form (for a detailed account see Rakic and Yakovlev 1968). The medial ganglionic eminence becomes recognizable at CS 14. At CS 15, the cerebral vesicles expand and are limited externally by the di-telencephalic sulcus. Both the lateral and medial ganglionic eminences can now be distinguished.

During the 6th week, at CS 16, the cortical wall can be subdivided into a future archipallium, paleopallium, and neopallium on the basis of histological features. According to Müller and O'Rahilly (1994), the prospective amygdaloid area is the most advanced structure displaying an outer primordial plexiform layer containing putative Cajal–Retzius cells. Olfactory fibers enter the brain at the site of the

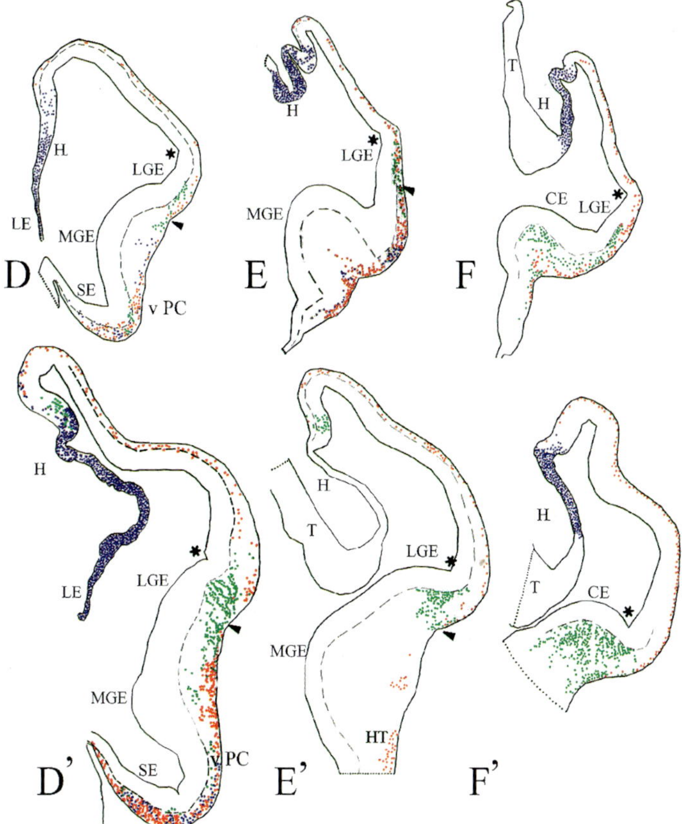

Fig. 1 (continued)

future olfactory bulb. At CS 17, the olfactory bulb appears, the hemispheres are growing, and the interhemispheric sulcus deepens. At the following stage, CS 18, a number of structures become recognizable, such as the anlage of the septum, placed between the olfactory bulb and the lamina terminalis, and a C-shaped cortical hem/hippocampal primordium, which extends all along the caudo-rostral extent of the hemisphere. The choroid plexus of the lateral ventricle begins to fold. In the following stages, CS 19 and CS 20 (48–51 days), the different structures become more and more prominent. On the whole, the period of cortical development from CS 14 to CS 20 corresponds to the preplate stage.

The defining feature of CS 21, at approximately 52 days, is the emergence of the first condensation of the cortical plate in the lateral aspect of the hemisphere. The cortical plate extends more medially in the next stage, CS 22, which also marks the end of the embryonic period.

In addition to the examination of classical Nissl-stained brain sections, the use of immunohistochemistry for the analysis of the expression of relevant brain molecules, for instance the calcium-binding proteins calbindin and calretinin and the extracellular matrix protein Reelin, allows a higher level of discrimination of the developmental events during the preplate stage.

In Fig. 1 we present a series of brain diagrams from rostral to caudal levels, obtained from parallel sections of human embryos at CS 17 (A–E) and 18 (A'–E'), which correspond to an initial preplate stage. It illustrates that early in telencephalic development, calretinin and calbindin are distributed in a complementary fashion. Calbindin dominates in the medial regions, including the "cortical hem," (see Sect. 6), septum, hippocampal primordium, and choroid plexus anlage, whereas calretinin is most prominent in the olfactory nerve and in the lateral pallium overlying the lateral ganglionic eminence. Close examination reveals a migratory stream of neurons from the lateral ganglionic eminence to the future paleocortex or olfactory cortex. The lateral ganglionic eminence is still rudimentary at CS 17 and few calretinin-positive cells have reached the paleocortex, while at CS 18 both ganglionic eminences are more prominent, in parallel with an increase of the calretinin-positive migratory pathway to the paleocortex. Reelin, in turn, is highly expressed in the basal forebrain and future septum, whereas only few Reelin-expressing Cajal–Retzius (CR) cells are visible in the prospective neocortex. Figure 1 represents the telencephalon prior to the complex radial and tangential migration waves that eventually lead to the establishment of the mature laminated cortex. Importantly, at this early stage an early regionalization of the embryonic telencephalon into medial, lateral, and ventral domains can be recognized. These domains will become less apparent after the formation of the cortical plate. Early regionalization has been studied extensively in the mouse on the basis of characteristic gene expression profiles (e.g., Bulfone et al. 1993; Bishop et al. 2000, 2002; Puelles and Rubenstein 1993; Ragsdale and Grove 2001), but thus far very little is known about their existence in the human brain (Lindsay et al. 2005).

3.2
The Early Foetal Period (8–20 GW): Time Course of Cell Proliferation and Migration

During the following period, from 8 GW to midgestation, huge numbers of neurons are generated in the proliferative zones, VZ and SVZ, and migrate into the cortical plate. According to Sidman and Rakic (1973), generation of neocortical neurons is highest during the first half of pregnancy and basically completed shortly after midgestation. However, cell proliferation and migration are not linear processes but undergo distinct phases. Based on the chemical analysis of DNA contents, Dobbing and Sands (1973) distinguished a first, rapid phase of cell growth from 10 to 18 GW, and a second, slower phase from approximately 20 GW until term. A more recent stereological study of cell numbers in the fetal human telencephalon similarly reported a fast, exponential increase from 13 to 20 GW and a slower, linear increase in from 22 GW to term (Samuelsen et al. 2003). Although these studies did not distinguish between neurogenesis, which predominates in the first half of gestation, and gliogenesis, which takes place mainly in the second half, they indicate that neurons are born and migrate also in the second half of gestation (and may continue at a slower pace during postnatal life). Information about the timing of neurogenesis is important because newborn and migrating neurons are particularly vulnerable to noxious influences.

3.3
Establishment of Cortical Lamination

The main feature of the cerebral cortex is the arrangement of its constituent neurons into horizontal layers, six in the neocortex and three in the archicortex or hippocampal formation. Each layer has a characteristic cell composition and connectivity. The local and regional differences in the laminar pattern, for example the prominence of the granular layers II and IV in the sensory cortices or the predominance of large pyramidal cells of layers III and V in the motor areas, constitute the basis of the unique variety of cytoarchitectonic areas of the adult human neocortex described in the classical neuroanatomical work of Brodmann (1909) and von Economo and Koskinas (1925).

The six-layered organization of the neocortex is established following a precise time table of neuronal birth, migration, and differentiation. Very little is known about this process in the human brain; most of our current knowledge is based on work performed in non-human species. According to the basic scheme of corticogenesis, established by birthdating studies in rodents and monkeys (Angevine and Sidman 1962; Berry and Rogers 1965; Rakic 1974), the deep cortical layers are the first to be born, to migrate, and to settle at their final destination. Later-born cohorts, destined to more superficial layers, successively migrate past the previous established layers and settle at the boundary between MZ and cortical plate. As a result, layer VI contains the oldest neurons and layer II the youngest neurons of the cortex. This "inside-out" gradient has been confirmed in all species examined, including monkeys (Rakic 1974). Ethical reasons prevent experimental birthdating studies in humans, but we may assume that the basic principles of "inside-out" lamination apply also to our species.

As a consequence of the inside-out principle, the fetal layers are formed not only by their definitive resident cells, but also by younger, still migratory neurons on their way to the top of the cortical plate, which are more abundant during the phases of intense neurogenesis. In his exhaustive review of human cortex development, Kostovic (1990a) distinguished between a first condensation of the cortical plate from 10 to 12 GW, the formation of a bilaminar cortical plate from 13 to 15 GW, a secondary condensation between 16 and 24 GW, and finally, around 32 GW, the gradual establishment of the definitive six neocortical layers. In the late fetal period, after 30 GW, the first cytoarchitectonic areas of Brodmann can be recognized. Nevertheless, the cortex of the newborn still shows signs of immaturity, such as a high cell density in layers II and IV, which differentiate after birth.

The sequence of neuronal maturation in the cortical layers is known from Golgi studies in human fetuses (e.g., Poljakow 1979; Marin-Padilla 1970; Mrzljak et al. 1988). Cajal–Retzius cells (see Sect. 7.3.1) in the MZ and subplate neurons (see Sect. 5.3.3) are the first neurons to achieve maturity, followed by large pyramidal cells in layer V and bitufted interneurons of layer III, which are recognizable at 26/27 GW. However, these neurons are an exception to the general rule of the deep-to-superficial gradient of dendritic differentiation that parallels the inside-out gradient of cell generation and migration. In Golgi-stained sections, the basic six layers of the neocortex can be recognized around 26–29 GW, although the upper layer III and layer II are still immature at term. Nonpyramidal neurons appear in all layers at the same time as pyramidal cells, although the maturation of the dendritic and axonal arborizations continues into postnatal life (Mrzljak et al. 1988; Poljakow 1979).

3.4
Development of Axonal Pathways

Unfortunately, our knowledge of the development of axonal connections in human brain is incomplete. Owing to ethical and technical difficulties, only few tracing studies have been performed in fetal human telencephalon. Electron microscope and acetylcholinesterase (AChE) histochemistry studies suggest that axons from the thalamus establish synaptic contacts first in the subplate by about 11–13 GW, and only later, by 23–25 GW, in the cortical plate (Kostovic and Molliver 1974; Molliver et al. 1973). Golgi studies by Marin-Padilla (1970) revealed the presence of 3 afferent horizontal axon plexuses in the cortex of a 7-months-old fetus. The most superficial plexus was in layer I, whereas the plexuses in layer VI and IV might represent the external and internal bands of Baillarger, respectively. Marin-Padilla (1970) emphasized the relative immaturity of the efferent cortical axons compared to the more advanced subcortical fiber systems in the intermediate zone.

Although injections of the lipophilic tracer DiI allow axonal tracing in post-mortem tissue, this method has rarely been used in human material. The few

DiI tracing studies suggest that the main fiber pathways of the cerebral cortex of humans develop similarly as in other mammals. DiI injections into the optic radiation showed that the majority of thalamocortical fibers terminate in the subplate at midgestation. As in other mammals, the long-projecting neurons in the cortical plate are in layers V, VI, and subplate (Hevner 2000). The intracortical axonal connectivity develops later in fetal life. In the visual cortex, vertical axonal projections between layers II/III and V are detectable at 26–29 GW, whereas horizontal connections appear later, at approximately 37 GW within layers IVB, V and VI (Burkhalter 1993; Burkhalter et al. 1993). The forward and feedback connections of visual areas V1 and V2 emerge shortly before birth, but do not acquire a mature appearance until 4 months of age (Burkhalter 1993).

The early appearance of the major fiber tracts can be also be recognized by examination of sections stained with adequate neurochemical markers (Fig. 2). The internal capsule appears as soon as the cortical plate is visible and pioneer cells in the subplate extend their axons (Fig. 2) as early as 8 GW. Figure 2 shows parallel sections from an 8-GW-old fetus where the major fiber tracts are just emerging.

Immunostaining for calbindin (Fig. 2A), calretinin (Fig. 2B), and doublecortin (Fig. 2C) shows a variety of fiber tracts shortly after the appearance of the cortical plate in the lateral aspect of the hemisphere. Calretinin-positive fibers originate from pioneer neurons in the subplate, which at this moment still reside within the cortical plate or begin to settle in a deeper position (Bayer and Altman 1990; Altman and Bayer 1991); they give rise to the first cortico–fugal system of the

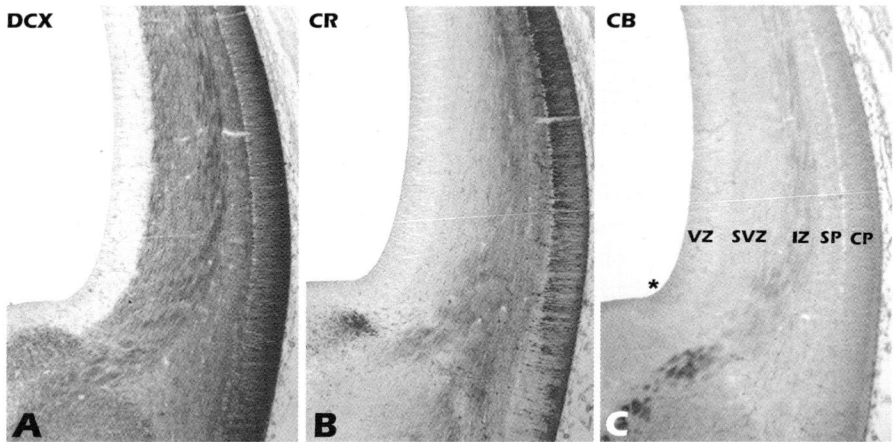

Fig. 2A–C Early fiber systems after the emergence of the cortical plate at CS 21. **A** Doublecortin (*DCX*) marks fibers in all compartments of the cortical wall except the VZ. **B** Calretinin is expressed in subplate neurons, early corticofugal fibers, and in cells of the lateral ganglionic eminence near the pallio-striatal angle (*asterisk*). **C** Calbindin marks thalamocortical fibers in the incipient internal capsule. *CP*, cortical plate; *IZ*, intermediate zone; *SVZ*, subventricular zone; *VZ*, ventricular zone

internal capsule. Calbindin-positive thalamo–cortical fibers have just arrived in the intermediate zone of the lateral wall, but have not yet entered the cortical plate. DCX is expressed in all fiber systems of the cortical wall and reveals a surprising number of fibers of unknown origin in the SVZ.

Pioneer fibers in the anterior commissure can be observed at 10 GW, the first fibers of the hippocampal commissure appear at 11 GW, and at 12–13 GW, the earliest fibers of the corpus callosum form within the "massa commissuralis," to grow soon and extend caudally (Rakic and Yakovlev 1968).

3.5
Neuronal Differentiation

Neurons start to differentiate once they have settled in their final position in the cortical plate. Pyramidal and nonpyramidal cells are present in all layers except in the molecular layer, which after the breakdown of the CR cell population contains only few small interneurons (Sect. 7.3.1.3). The immense diversity of dendritic and axonal branching patterns of nonpyramidal neurons has been described in earlier Golgi studies in the cortex of rodents, carnivores, and humans (Cajal 1911; Fairen et al. 1984; Meyer 1983, 1987; Meyer and Ferres-Torres 1984; Poljakow 1979), and clearly shows a trend toward an increasing complexity in neuronal morphology and architecture. The description of the impressive variety of morphologically and neurochemically distinct interneurons in the human cortex is beyond the scope of this work, the more so as this aspect of cortical differentiation has rarely been addressed from a developmental point of view.

Detailed Golgi studies in the fetal prefrontal cortex illustrate the initial immaturity of the cortical plate compared to the precocious maturity of the MZ and the subplate, and provide a timetable of neuronal differentiation (Mrzljak et al. 1988). At 10/12 GW, neurons in the superficial cortical plate display a bipolar shape with a leading process that branches extensively in the MZ, and a fine descending process that reaches into the intermediate zone. Neurons in the deeper cortical plate appear more differentiated. The first basal dendrites of pyramidal cells develop by 15 GW and extend secondary branches from 17–25 GW. By 26–29 GW, the first dendritic spines appear on the proximal dendritic segments of apical and basal dendrites of pyramidal cells of layers V and III. Concurrently, a type of cortical interneurons, the double-bouquet cells, extend columnar dendritic and axonal trees. At birth, pyramidal cells have not yet reached full maturity, and especially the cells of layer II have poorly developed dendritic trees. Interneurons of the basket cell variety are present in layer V, but do not yet display the characteristic pericellular axonal varicosities. Similar observations were made by Marin-Padilla (1970) in the motor cortex. The first dendritic spines appeared on pyramidal cells of layer V, whereas the characteristic axonal terminals of basket cells appeared postnatally.

Basket cells and axo–axonic interneurons or "chandelier cells" (Somogyi et al. 1982; Szentagothai 1978) are important types of GABAergic inhibitory neurons that express the calcium-binding protein parvalbumin (reviewed by Howard et al.

2005). The first parvalbumin-positive neurons appeared in the cortical plate at 26 GW in layer VI, spreading into more superficial layers following the general inside-out sequence of cortical development (Honig et al. 1996). The characteristic axonal chandelier terminals have not been reported prenatally, and seem to be a postnatal feature (Mrzljak et al. 1990; own unpublished observations).

3.6
Cortical Gyration

The adult human cortex is highly folded. Remarkably, gyration and sulcation patterns are not evident until after midgestation, when neurogenesis and migration are almost completed and areal architecture and laminar arrangement become recognizable. Gyration proceeds following a precise sequence which provides important landmarks for the evaluation of the gestational age and the diagnosis of possible malformations using magnetic resonance imaging (MRI) techniques (e.g., Levine and Barnes 1999; Garel et al. 2001). The first fissures to appear are the primary fissures: the Sylvian, calcarine, and cingulate fissures are recognizable by 16 GW, the parieto-occipital fissure at 18 GW, the Rolandic or central fissure at 20/21 GW. The secondary fissures—in the temporal lobe, these would be the superior temporal, second transverse, and collateral fissures—appear later, after 30 GW, as side branches of the primary fissures, and further branching will give rise to the tertiary fissures, which become apparent at term (Gilles and Gomez 2005). It is important to note that the timetable of gyration indicated here is obtained from anatomical studies or atlases of human brain development (Sidman and Rakic 1982; Chi et al. 1977; Fees-Higgins and Larroche 1987; Bayer and Altman 2004) and precedes the timetables observed in imaging studies (Garel et al. 2001). Some primary fissures, in particular the Sylvian fissure and the hippocampal fissure, appear even earlier than described above, since at close inspection they can be detected as shallow grooves already by 12 GW. These fissures are very constant even in lissencephalies caused by mutations of certain genes, when the tertiary and secondary sulci are missing (see Sect. 9).

Cortical gyration is usually attributed to mechanical constraint: the evolutionary increasing cortical surface has to adapt to a restricted space in the cranium and thus begins to fold. However, highly folded neocortices are found not only in primates, but also in large-brained representatives of different lineages, as exemplified by the brains of two monotremes, the large gyrencephalic echidna and the small lissencephalic platypus. The degree of cortical gyration seems to correlate more strongly with absolute neocortex size than with taxonomic affiliation, suggesting that folding is causally related to brain size. In any case, the mechanical constraint theory does not explain why cortical width is so constant across mammals, which indicates that cortical gyration is also influenced by other as yet unknown mechanisms (Strieter 2005).

4
Neuronal Migration

Postmitotic neurons have to leave their birthplace in the proliferative compartments and migrate to their final destination in the cortical plate. The migratory route is relatively short during the preplate stage and then progressively increases, reaching several millimeters by midgestation. Neuronal migration is a complex process that involves tangential and radial movements under tight genetic control. A variety of migration disorders in humans and in mutant mice are due to mutations of key genes involved in diverse aspects of migration, some of which will be described in detail in Sect. 9.

Gliophilic radial migration has long been considered as the only mode of cortical migration. Cortical neurons were thought to be generated exclusively in the proliferative zones of the cortical wall and to migrate into the cortical plate using radial glia as guides (Rakic 1972). Recent evidence in rodents has shown that this concept is valid only for the excitatory pyramidal neurons, whereas GABAergic inhibitory neurons migrate tangentially from their origin in the ganglionic eminences, the primordium of the basal ganglia (Andersen et al. 1997, 1999, 2002; Lavdas et al. 1999; Parnavelas 2000; reviewed by Marin and Rubenstein 2001). Projection neurons and interneurons thus have different origins and use different migration modes.

4.1
Radial Migration

The glutamatergic pyramidal cells are the main projection neurons and express the transcription factors Emx1 and Tbr1; they are born in the cortical VZ (Chan et al. 2001; Hevner et al. 2001; Miyata et al. 2001; Noctor et al. 2001; Gorski et al. 2002; Hatanaka and Murakami 2002; Weissman et al. 2003) and migrate radially to the cortical plate.

Somal translocation is a distinct mode of radial migration that predominates in the early stages of cortical development (Nadarajah et al. 2001, 2002). The earlier-generated neurons undergo somal translocation from the VZ to a position beneath the pial surface. They typically have a long, radially oriented, leading process that terminates at the pial surface (Fig. 3A, C), and a short, transient trailing process (Fig. 3B). Translocating cells propel their nucleus toward their leading process, that is stably attached to the pia, while the trailing process detaches from the ventricle. They move through a continuous advancement that allows fast migration. This mode of migration is distinct from glia-guided or gliophilic locomotion characteristic of later-born cortical-plate cells. Locomoting cells have a shorter radial process that is not attached to the pial surface, and they move with a slow, saltatory pattern with intermittent resting periods that result in slower migration speed. However, at the end of migration, when their leading process reaches the MZ, locomoting cells seem to switch from the saltatory movement to translocation.

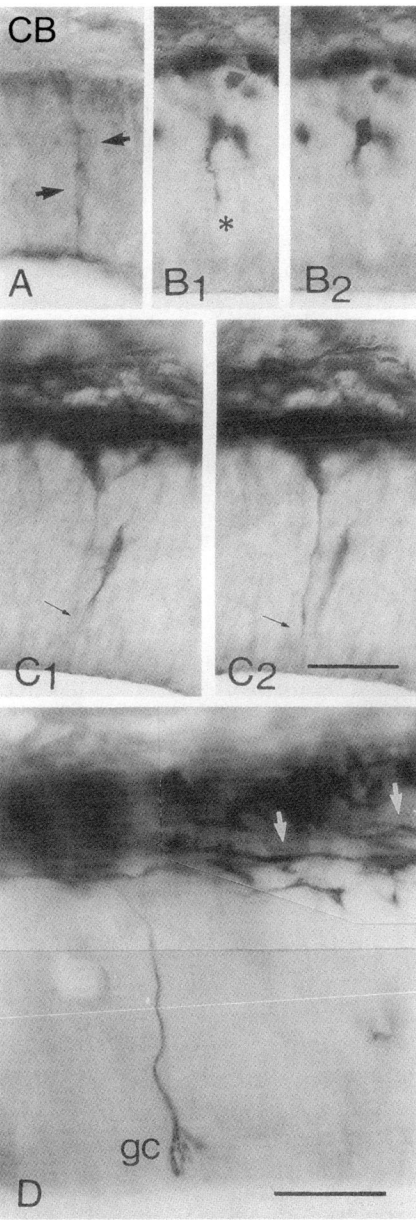

Fig. 3A–D Calbindin (CB)-positive pioneer cells in the rat cortex at E11, displaying the morphology of neurons migrating by somal translocation. In **A** the *arrows* point to the soma and a process extending to the pial surface. **B₁** and **B₂** Different focal points of a neuron that has retracted its ventricular process. In **C₁** and **C₂** Neurons connected to both the VZ and pial surface. **D** Early pioneer fibers extending growth cones into the VZ. (From Meyer et al. 1998, with permission)

Radial locomotion may be more complex than previously believed. Recent studies indicate that pyramidal cells may not simply attach to a radial glial fiber and then travel to the cortex, but that radial migration can be subdivided into at least four different phases (Noctor et al. 2004; Kriegstein and Noctor 2004). In the first phase, after the classic interkinetic nuclear migration and cell division at the ventricular surface, postmitotic neurons ascend radially from the VZ to the SVZ. In a second phase, they pause, or sojourn, in the SVZ and take a multipolar shape (Tabata and Nakajima 2003, Noctor et al. 2004). After about one day, they convert into a bipolar shape and resume migration. They can now move tangentially in the SVZ without being attached to a radial glial fiber and then take different destinations. Some neurons may go through a third phase and move back to the VZ, and in a fourth phase, reverse polarity and migrate radially to the cortical plate. Other neurons may pass from the second phase, the sojourning in the SVZ, directly to phase 4, the radial migration into the cortical plate. This very complex migratory behavior of postmitotic neurons makes it more difficult to analyze human migration disorders. Different genes may be involved in the control of specific phases of migration and their progression: For example, an arrest in phase 2 may lead to an impossibility to change to the radial mode and migrate into the cortical plate, giving rise to a potential migration disorder. In fact, *LIS1*-deficient cells have been shown to be unable to progress from the multipolar non-migratory state to the bipolar migratory state (Tsai et al. 2005; see Sect. 9.3). Another, distinct abnormal mode of radial migration, branched migration, has been described in *p35*-deficient mice. In contrast to normal glia-guided locomoting cells, which possess an unbranched leading process, migrating *p35-/-* cells show extensive branching of the leading process and no association with a radial glia fiber (Gupta et al. 2003).

4.2
Tangential Migration

Cortical interneurons are GABAergic and express the transcription factors Dlx, Nkx2.1 and Lhx6; in rodents and ferrets they are born in the medial, lateral, and caudal ganglionic eminences from where they migrate tangentially into the cortex (Anderson et al. 1997, 1999, 2001, 2002; DeCarlos 1996; Tamamaki et al. 1997; Lavdas et al. 1999; Sussel et al. 1999; Parnavelas 2000; Corbin et al. 2001; Wichterle et al. 1999, 2001, 2003; Ang et al. 2003). Interneurons use various tangential migratory routes to reach the cortex, traveling through the lower IZ, SVZ, MZ, and subplate (Anderson et al. 1997; Lavdas et al. 1999; Polleux et al. 2002; Ang et al. 2003; Tanaka et al. 2003; Hevner et al. 2004). Interneurons migrating through the MZ originate from different sources and form distinct and independent streams before they enter the cortical plate from above and descend to assume their position side by side with isochronically generated pyramidal cells (Ang et al. 2003). Interneurons migrating though the SVZ and IZ turn upward and enter the cortex from below, a movement that may be guided by radial glia (Polleux et al. 2002). The guidance substrate of tangential migration is not known.

Mutant mice with genetic deletions of Dlx1 and Dlx2 have shown that both genes are required for the migration of interneurons from the ventral telencephalon to the cortex (Anderson et al. 1997; 1999). The neuronal adhesion molecule TAG-1, expressed on developing corticofugal axons, mediates this process (Denaxa et al. 2001). Time-lapse imaging of acute brain slices showed that GABAergic interneurons do not take the most direct route into the cortex but first actively direct toward the VZ before they assume a radial course into the cortical plate, perhaps to acquire layer information (Nadarajah et al. 2002). According to Tanaka et al. (2003), cortical GABAergic neurons display complex, multiple modes of migration: an ordered migration in a ventrolateral-to-dorsomedial direction along the lower IZ and the subventricular zone, a radial and non-radial migration toward the pial surface, a multidirectional migration in the tangential plane of the marginal zone (MZ), and a radial migration from the MZ to the cortical plate.

GABAergic interneurons do not belong to a homogeneous cell class but display an immense morphological and neurochemical diversity. Non-overlapping subgroups of interneurons in the rodent cortex express parvalbumin, somatostatin, and calretinin (Rogers 1992; Kubota et al. 1994), and together comprise more than 80% of interneurons (Gonchar and Burkhalter 1997). Interestingly, different classes of interneurons originate from distinct subcortical sources. Parvalbumin and somatostatin-expressing interneurons originate primarily within the medial ganglionic eminence, whereas calretinin-positive interneurons seem to derive mainly from the caudal ganglionic eminence (Xu et al. 2003, 2004). The caudal ganglionic eminence provides interneurons for caudal cortical areas, hippocampus, and some amygdalar nuclei (Nery et al. 2002). It is generally accepted that in rodents the immense majority of interneurons derive from subcortical sources, which is in contrast with observations in the human cortex, where large numbers of interneurons are generated in the cortical SVZ (Letinic et al. 2002; Rakic and Zecevic 2003).

The molecular guidance mechanisms that regulate tangential and radial migrations may be different. For instance, the chemokine SDF-1 is a potent chemoattractant for migrating interneurons, but not for pyramidal cells, and deficiency of SDF1 or its receptor CXCR4 gives rise to a migrational defect affecting interneurons of the upper layers but not pyramidal neurons (Stumm et al. 2003). A variety of factors have been shown to participate in the guidance of tangentially migrating interneurons, such as semaphorin–neuropilins (Marin et al. 2001; Tamamaki et al. 2003), polysialylated neural cell adhesion molecule (PSA-NCAM), neuregulins (Yau et al. 2003), and the slit and robo families of attractant and repellent molecules (Zhu et al. 1999). For instance, Slit is expressed in the VZ of the ganglionic eminences and repels SVZ cells that leave the ganglionic eminence and settle in the striatum or migrate to the cortex (Zhu et al. 1999). In turn, semaphorins are expressed in the striatum and repel neurons that express the neuropilin receptor, causing these neurons to avoid the stratum and direct toward the cortex (Marin et al. 2001; Marin and Rubenstein 2001).

Despite the different origins and migration modes of interneurons and pyramidal neurons, their eventual laminar fates are highly regulated and coordinated. Both neuronal classes are arranged according to the same "inside-out" gradient that correlates laminar fate and day of birth (Angevine and Sidman 1961, Fairen et al. 1986; Miller 1986; Peduzzi 1988). The laminar positioning of interneurons is an important issue also for the human cortex, since abnormal distribution of interneurons has been described as associated with epilepsy (Hannan et al. 1999; Thom et al. 2004). In humans, malpositioning of interneurons may be so severe as to cause so-called "interneuronopathy," a term coined for X-linked lissencephaly with abnormal genitalia and West's syndrome associated with *ARX* mutations. The interneuronopathy is considered to be caused by a disorder affecting the tangential migration of interneurons (Kato y Dobyns 2005).

4.3
Kinetics of Neuronal Migration

Both radially and tangentially migrating neurons apparently use the same basic mechanisms described for motile cells in non-neural tissues. Three main events have been distinguished in the process of migration: first, the extension of a leading process, second, a displacement of the nucleus or nucleokinesis, and third, the retraction of the trailing process at the cell's rear (Morris et al. 1998; Walsh and Goffinet 2000). When a cell initiates migration, it becomes polarized and extends a leading edge, a process oriented in the direction of migration that terminates in a ruffling lamellipodium, and has a narrow, retracting tail at the rear. Leading-edge extension involves a dynamic assembling and disassembling of filopodia and lamellopodia, which respond to attractive or repulsive guidance cues and allow the cell to sample the environment before initiating somal migration. Time-lapse imaging has shown that nucleokinesis is preceded by a dilatation of the leading process accompanied by a forward translocation of the centrosome and Golgi apparatus. The nucleus then jumps into the dilatation in a saltatory movement. Nonmuscle myosin II seems to accumulate periodically behind the nucleus and may contract the rear of the migrating neuron to push the nucleus forward, so that actinmyosin contraction provides the driving force for nuclear migration (Bellion et al. 2005; Schaar and McConnell 2005). The dynamics of the actin cytoskeleton is regulated by small GTPases of the Rho family and closely interconnected with the microtubule cytoskeleton (reviewed by Luo 2000; Wittmann and Waterman-Storer 2001).

5
Compartments of the Cortical Wall

The developing cortical wall is organized into transient strata (Boulder Committee 1970) which continuously change as neurons migrate and fiber tracts develop. Figure 4 shows the changes in cortical stratification from the early preplate to

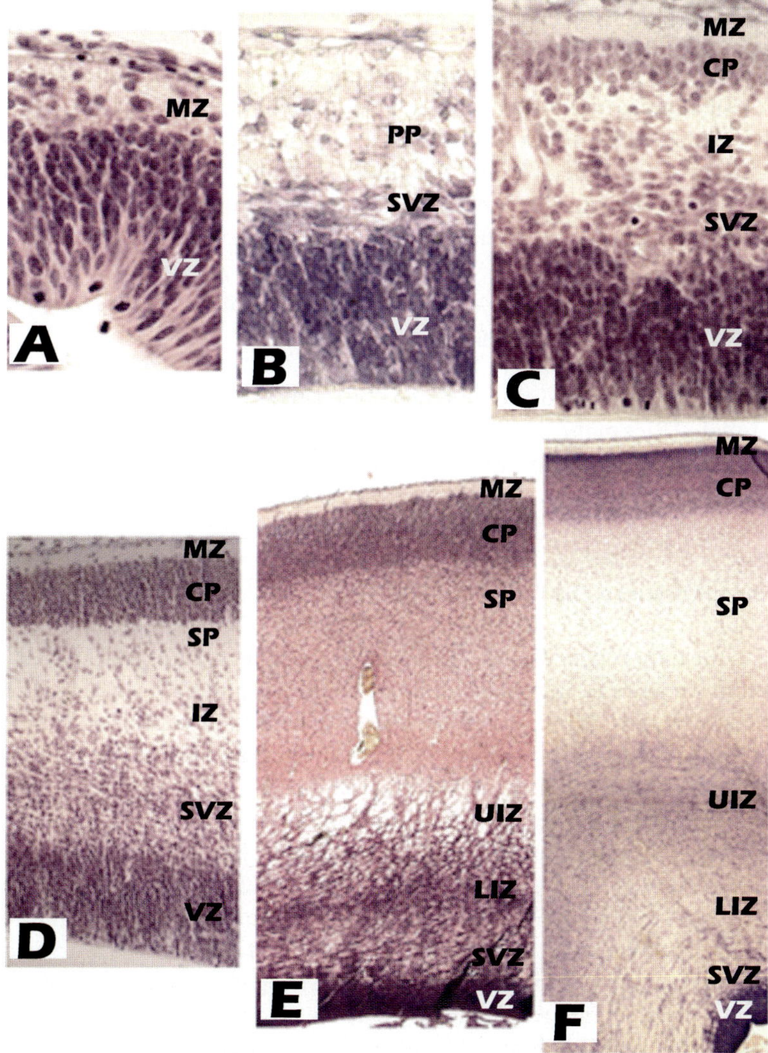

Fig. 4A–F Compartments of the cortical wall from 6 to 22 GW stained with cresyl violet. A Early preplate stage, 6 GW (CS 17). B Advanced preplate stage, 7 GW (CS 19). C The initial cortical plate (or pioneer plate) stage, 8 GW (CS 21). D Layering of the cortical wall at 9 GW. E 13 GW. F 22 GW. Note the size increase of subplate and IZ from 9 to 22 GW

the highly enlarged cortical wall at 22 GW. We have represented the following main developmental steps: The early preplate at 6 GW (Fig. 4A) consists of a thick proliferative ventricular zone (VZ) and a narrow, cell-sparse marginal zone (MZ). In the advanced preplate stage at 7 GW (Fig. 4B), an additional proliferative zone, the subventricular zone (SVZ) has appeared at the outer surface of the VZ, and the MZ is now occupied by numerous, mainly horizontally oriented neurons. The initial cortical plate (ICP) or pioneer plate stage at 7.5/8 GW (Fig. 4C) is characterized by the first condensation of radially organized neurons that form a dense cell aggregation below a cell-sparse MZ. In successive stages (Fig. 4D–F) the cortical plate increases in width, and the migratory compartments—lower and upper intermediate zone (IZ) and subplate—develop and differentiate, reaching dimensions that are characteristic of the human or primate brain, far beyond their counterparts in the rodent brain.

5.1
Compartments at the Preplate Stage

The preplate is considered the first stage of cortical development, preceding the cortical plate (Rickmann and Wolff 1981; Rickmann et al. 1981; De Carlos and O'Leary 1992). It extends between the pia and the pseudostratified proliferative neuroepithelium of the VZ that lines the lateral ventricle.

5.1.1
The Neuroepithelium of the Ventricular Zone

The VZ is the only germinal zone until the SVZ appears at 7 GW (Chan et al. 2002). Most of our knowledge on cell proliferation in the VZ and SVZ comes from rodent and monkey studies, although proliferation markers such as proliferating nuclear antigen (PCNA) have also been used to study human embryonic development (Chan et al. 2002).

In the embryonic cortex, cells in the VZ undergo a movement known as "interkinetic nuclear migration": during the G1 phase, nuclei move away from the apical ventricular surface to the basal region of the VZ where they stay during the S-phase. In G2, they return to the apical region and mitosis occurs at the ventricular surface (Takahashi et al. 1993, 1995). Early multipotential precursor cells in the VZ give rise to progenitors with a more restricted neuronal or glial progeny. Cell lineage experiments revealed the existence of diverse sets of precursor cells specified for the generation of specific cell classes (Luskin et al. 1988; Grove et al. 1993). Initially, all neuroepithelial cells have the same morphology, spanning the entire width of the wall of the cortical primordium. They are coupled together into clusters by gap junctions, channels that may allow direct cell-to-cell interactions (LoTurco et al. 1991; Mollgard et al. 1987). The clusters contain only radial glia and neuronal precursors, and increased uncoupling in S phase during late neurogenesis may contribute to the greater percentage of VZ cells exiting the cell cycle (Bittman et al. 1997).

After the onset of neurogenesis, the neuroepithelial cells differentiate into radial glia cells, which are attached to the ventricular surface by a short process, and which extend a long radial process that terminates at the pial surface. The radial glia is a multipotential progenitor cell class that gives rise to neurons and astrocytes of the embryonic brain (Hartfuss et al. 2001; Malatesta et al. 2000; Miyata et al. 2001; Noctor et al. 2001, 2002; Tamamaki et al. 2001; Parnavelas and Nadarajah 2001). Importantly, even at the early human preplate stage (5–6 GW), the radial glia is not homogeneous but expresses multiple sets of neuronal and glial marker molecules suggesting the coexistence of multipotent neuroepithelial cells, radial glia, and restricted neuronal progenitors (Zecevic 2004).

Neurogenesis proceeds through a combination of several modes of division: symmetric progenitor cell divisions expand the pool of precursor cells, while asymmetric progenitor cell divisions give rise to one self-renewing radial glia cell and a single neuron. Symmetric terminal divisions take place in the SVZ (Noctor et al. 2004; Haubensak et al. 2004) and consist in the production of two daughter neurons, thus depleting the pool of precursor cells. The progression from predominantly symmetric to predominantly asymmetric divisions may be influenced by the transcription factor Pax6 (Estivill-Torrus et al. 2002), and the transition from radial glia to progenitor cells is associated with a downregulation of Pax6 and upregulation of Tbr2, a T-domain transcription factor (Englund et al. 2005). It has been proposed that early changes in the ratio of symmetric proliferative divisions versus asymmetric neuronogenic divisions might influence cortical size, such that an increase of the production of progenitor cells would lead to an increase in the total number of differentiated neurons (Caviness et al. 1995; Rakic 1995). Since the human cortex is characterized by a huge expansion of the cortical surface area, it would be interesting to know which mode of cell division prevails at the early stages of human development and which genes control the different aspects of proliferation modes. According to Chan et al. (2002), asymmetric cell division is the predominant mode of proliferation after the age of 12 GW, while remnants of the proliferative SVZ persist into adulthood (Alvarez-Buylla and Garcia-Verdugo 2002; Gage 2000, Merkle et al. 2004).

Interestingly, the cell-cycle duration of cortical precursor cells is more prolonged in primates than in rodents, and it has been hypothesized that the prolongation of the cell cycle in cortical precursors might be an adaptive feature underlying the evolutionary expansion of the primate neocortex (Rakic 1995).

5.1.2
The Preplate

The prevailing preplate concept is based on the "plexiform primordial layer," which was postulated to be a common feature of all embryonic mammals as a representative of a hypothetical ancestral stratum present in amphibians and reptiles (Marin-Padilla 1971, 1972, 1978, 1990). According to this hypothesis, the preplate contains the oldest neurons of the cortex, namely CR cells and subplate or layer VII cells, both of which were believed to originate from the local neuroepithelium. In Golgi sections, they display distinct morphologies: CR cells are described as being horizontal and bipolar, whereas subplate neurons are multipolar or pyramidal-like. The appearance of the cortical plate splits the two components of the preplate: CR cells settle in the MZ, whereas subplate neurons reside below the cortical plate in the future white matter, and thus form a framework for the developing cortical plate (Marin-Padilla 1978). This process, known as the partition or splitting of the preplate, is considered a key event in corticogenesis and the main paradigm to explain early cortical malformations in transgenic mice and human gene mutations. The basic fact, the formation of a preplate and the subsequent insertion of the cortical plate within the preplate has been confirmed in numerous birthdating experiments in a variety of mammals (Bayer and Altman 1991; Luskin and Shatz 1985; Raedler and Raedler 1978; Wood et al. 1992; Sheppard and Pearlman 1997).

Whereas CR cells are widely believed to be the dominant cell class of the preplate, more recent studies indicate that even in the rat there are additional neuronal types that complicate the simple scheme of a preplate splitting into CR and subplate cells.

A prominent cell class, the superficial pioneer cells, are born in the rat at E 11 and appear concurrently with Reelin-positive CR cells. Pioneer cells express calretinin or calbindin and are Reelin negative. They remain in the MZ/upper cortical plate after the emergence of the cortical plate (Meyer et al. 1998), whereas deep pioneer cells come to reside below the cortical plate and form part of the subplate (Fig. 5A–D). Interestingly, some preplate pioneer cells were reported to derive from the medial ganglionic eminence and to reach the preplate by tangential migration (Morante-Oria et al. 2003).

Calbindin- and calretinin-expressing pioneer cells first extend transient fibers that descend into the VZ and disappear at E 14 (Fig. 3D). At E 16, axons originating from superficial pioneer cells traverse the cortical plate, branch in the subplate, and may even send fibers into the nascent internal capsule (Fig. 5A–C; Meyer et al. 1998; Soria and Fairen 2000). The axonal preplate projection into the internal capsule is usually attributed to the subplate component (McConnell et al. 1989; De Carlos and O'Leary 1992), rather than to the superficial pioneer neurons. The fate of the superficial pioneer cell population of the MZ is unknown, although BrdU data in rats suggest that they disappear before birth (Meyer et al. 1998). The deep pioneer population belongs to the subplate, which in the rat becomes prominent at E 16 with calbindin (Fig. 5D). The morphology and axonal projections of superficial and deep pioneer neurons, as well as their relationship with the incipient cortical plate, have been beautifully illustrated in the Golgi studies of the cat cerebral cortex by Marin-Padilla (1978; Fig. 4), where they were identified as CR cells. This initial misinterpretation may have contributed to the overestimation of the role of the Reelin-positive CR cells in preplate splitting.

5.2
The Transition from Human Preplate to Cortical Plate

We examined the main events in preplate development in the embryonic human cortex and found evidence for a succession of transient events that gradually lead to the condensation of the cortical plate. These events differ from those outlined above in the rat, where CR cells and pioneer cells appear almost simultaneously.

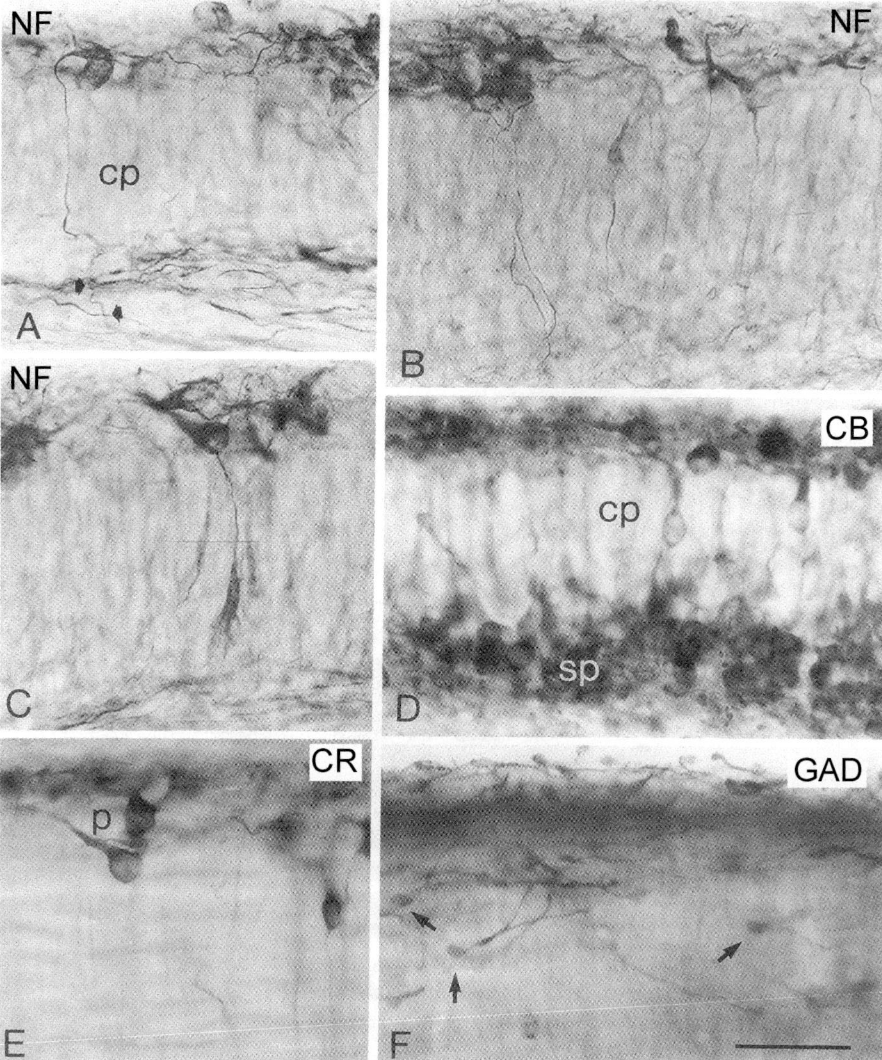

Fig. 5A–F Pioneer cells in the rat cortex at E 16, stained with neurofilament (*NF*) antibodies in **A–C** and calbindin (*CB*) in **D**. Pioneer cells often form clusters in the MZ and send descending fibers into the subplate and incipient internal capsule, traversing the cortical plate. Deep pioneer cells form part of the subplate (**D**). **E** E 18, calretinin (*CR*)-positive pioneer cells. **F** E 16, GAD-positive subpial granule cells in the MZ. (From Meyer et al. 1998, with permission)

We found that human Reelin-expressing CR cells develop gradually and that various CR cell populations emerge successively during the first 3 months of gestation (discussed in detail in Sect. 7.3.1). At 5 GW, (CS 16; Fig. 6A), there was only a neuroepithelium and a narrow, cell-sparse marginal layer, and the few first Reelin-positive neurons in the prospective neocortical territory were scattered below the pial surface. Reelin-positive cells increased in numbers from CS 15 to CS 18 (6.5 GW; Fig. 6B–D), when they formed a continuous row in the still narrow preplate. At CS 17, no other neuronal classes were observed, and thus Reelin-positive neurons were in fact the first neurons of the prospective neocortex. The first Reelin-negative cell populations appeared in the preplate at CS 19. A few calretinin-positive, Reelin-negative cells assumed a horizontal orientation, while others lay in the superficial VZ or occasionally in the deep VZ (Meyer et al. 2000).

The continuous appearance of new cell types from 6 GW to 8 GW led to an increasing complexity that is already evident in Nissl-stained sections (compare Fig. 4A, early preplate and 4B, advanced preplate). Using antibodies against calretinin, Reelin and GAD, we distinguished several transitional steps that eventually lead to the formation of the cortical plate.

In the advanced preplate at 7 GW (CS 20), CR cells were more numerous and confined to a position just below the pial surface. At this stage, large neurons belonging to a new cell class arrived in the preplate and settled in a single layer external to the SVZ, which also appeared at this stage. These large neurons were bipolar and appeared morphologically more differentiated than CR cells. They expressed GAD and calretinin and were Reelin-negative (Fig. 7A–D). Because of their arrangement in a single cell layer we termed them "monolayer cells."

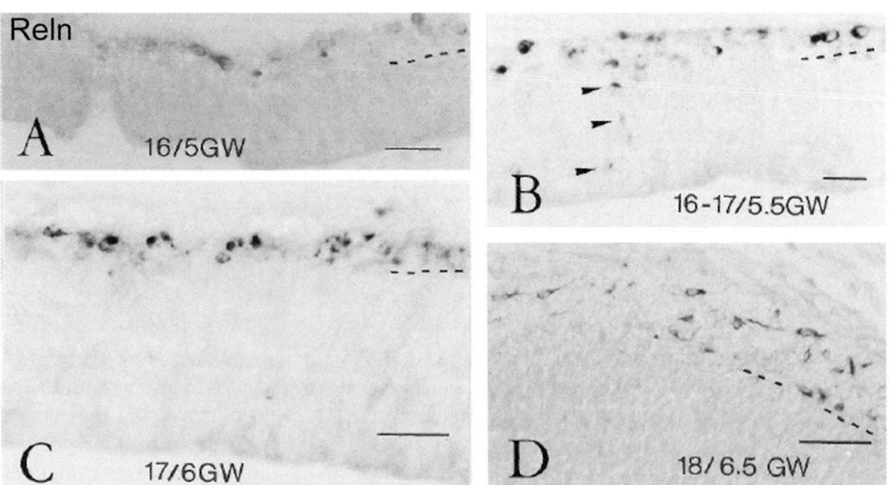

Fig. 6A–D CR cells in the early stages of corticogenesis. **A** 5 GW (CS 16). The first Reelin (*Reln*)-positive CR cells appear at the outer prospect of the cortical neuroepithelium. **B** 5.5 GW (CS 16/17) in the marginal layer. The deep cells in the VZ suggest a local origin. **C** 6 GW (CS 17). **D** 6.5 GW (CS 18), CR cells slowly increase in number but remain sparse compared to later stages, indicating a protracted initial phase of corticogenesis. (Adapted from Meyer et al. 2000, with permission)

The presence of GABAergic neurons is usually not included in the descriptions
of preplate cell populations, although they were also observed by Zecevic and
Milosevic (1997) above and below the incipient cortical plate of human embryos.
The example of the "monolayer cells" clearly demonstrates that the identification
of a neuronal type cannot rely on morphology alone, because in Golgi-stained
sections they would probably be confounded with CR cells. The neurochemical
phenotype is thus as important for defining a neuronal class as its morphology.

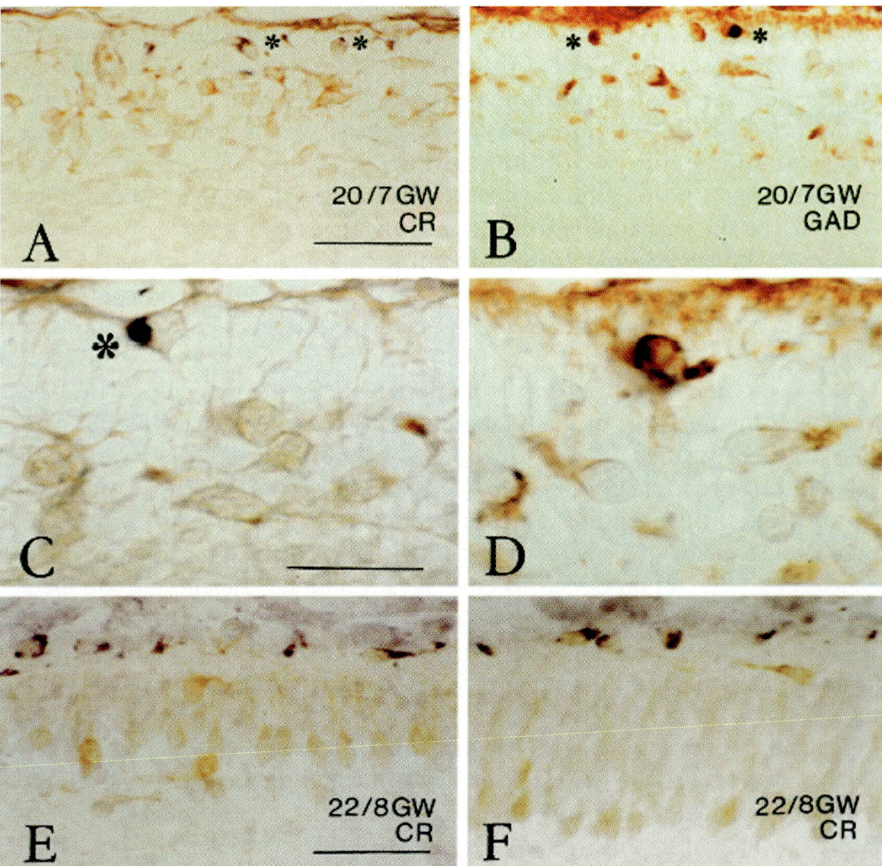

Fig. 7A–F Cell populations in the advanced preplate and early cortical plate. **A–D** 7 GW
(CS 20). **A** and **C** are double-labeled with Reelin (*black, asterisks*) and calretinin (*light
brown*). **B** and **D** show Reelin (*black, asterisk* in B) in small CR-cells close to the pial surface.
Horizontal neurons in the monolayer express CR and/or GAD but are Reelin negative.
There is thus a clear segregation of Reelin-positive CR cells and the other, Reelin-negative
components of the preplate. **E** At CS 21, the calretinin-positive initial cortical plate or
pioneer plate forms below the monolayer, whereas CR cells are confined to their subpial
compartment. **F** From the same case, at a more ventral level, showing calretinin-negative
cells in the more advanced cortical plate. (From Meyer et al. 2000, with permission)

CS 21 (7 GW) marks a new step in the progression from the preplate to the cortical plate. Many calretinin-positive, GAD-negative neurons begin to aggregate below the GAD-positive monolayer without any preferential orientation, being more numerous in the lateral cortex than in the dorsal cortex (compare Fig. 8A–C with 8D–F). In turn, Reelin-positive CR cells occupied the most superficial part of the preplate just below the pia, where they remained during the following weeks. In the next step, at CS 22 (8 GW), the densely aggregated calretinin-positive neurons, which we termed "pioneer neurons" because of their similarity to the pioneer cells of the rat, assumed a radial orientation and emitted the first axonal fibers from the deep pole of their somata (Fig. 8G). They thus represent the first projection neurons of the cortex. In successive steps (which can be observed in a single section of this stage, comparing the less differentiated dorsal cortex with the more advanced lateral cortex, Fig. 8G–J), this initial "pioneer plate" was invaded by an increasing number of calretinin-negative neurons, which represented the first cohorts of the cortical plate and inserted themselves between the superficial and the deep pioneer neurons. The superficial pioneer neurons became sparse, probably because they belong to a limited cell population, soon dispersed by the growing cortical plate, whereas the deep pioneer cells assumed a pyramid-like shape and continued to increase in number (Fig. 8J), forming the subplate anlage, or "presubplate" in the terminology of Kostovic and Rakic (1990).

This event, the "splitting" of the pioneer plate by the arriving cortical-plate neurons, is the final step of the transition from the preplate stage to the cortical plate stage, represented schematically in Fig. 9. It differs from the concept of pre-plate splitting proposed by Marin-Padilla (1978) insofar as CR cells are apparently not involved in this step but remain from the beginning in their own separate compartment below the pial surface and are at no time point intermixed with pioneer neurons. Our observations that CR cells migrate tangentially into the MZ (Meyer and Wahle 1999; Meyer et al. 2002a; see also Muzio and Mallamaci 2005; Yoshida et al. 2006) lend support to the hypothesis that preplate partition involves primarily superficial and deep pioneer neurons rather than CR cells and subplate cells.

The emergence of the neocortical plate marks the end of the embryonic period and the onset of the fetal period. From this moment, the main three cortical strata, the marginal zone, the cortical plate, and the subplate, will increase in size and become populated by the myriad migrating neurons that travel along radial and tangential pathways, until the MZ and the subplate undergo regressive changes by the end of corticogenesis.

The developmental steps outlined here can be distinguished only in series of human or monkey embryos of closely consecutive ages. In rodent embryos, the low temporo-spatial resolution makes it difficult to clearly dissect the multiple cellular events that take place in the course of approximately two weeks of human embryonic life. Nevertheless, a comparison of the human data presented here with those in the rat (Meyer et al. 1998) suggests different time courses in the appearance of the main preplate components, with a dominance of CR cells in the

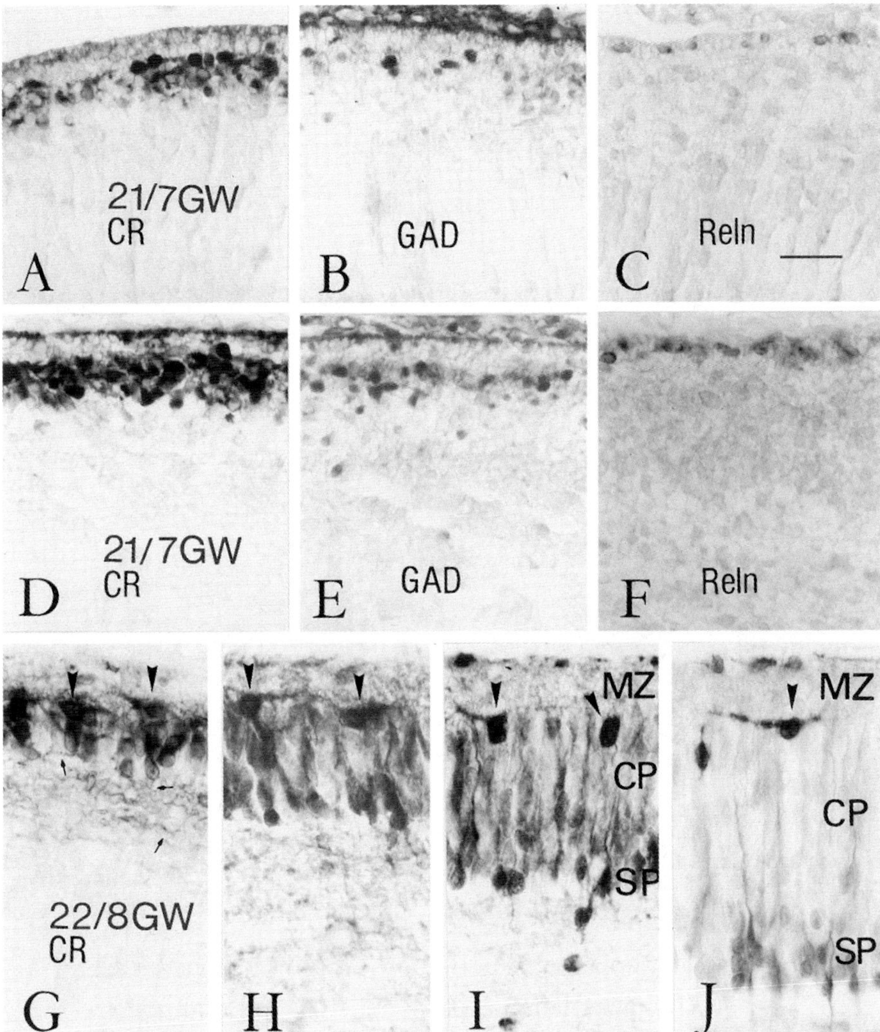

Fig. 8A–J Partition of the pioneer plate and establishment of the cortical plate. A–F 7 GW (CS 21). A–C Sections from the more immature dorsal cortex, stained for calretinin (A), GAD (B) and Reelin (C). The intensely calretinin-positive pioneer plate is an aggregate of horizontal and more rounded cells (A), which contains in its upper part a few GAD-positive neurons (B). CR cells are immediately below the pia (C). D–F are from the more mature lateral cortex. The pioneer plate is thicker (A), GAD-positive neurons are more numerous and concentrated in the superficial pioneer plate (E), whereas CR cells remain in their subpial location (F). G–J: Stage 22, 8/9 GW, calretinin. There is a gradual splitting of the pioneer plate from the most dorsal (G) to intermediate (H, I) and more lateral (J) levels. The intensely calretinin-positive pioneer plate is progressively invaded by calretinin-negative cortical plate cells that insert themselves between the superficial and deep pioneer neurons. The superficial pioneer population seems to disappear, because it is not seen at later stages

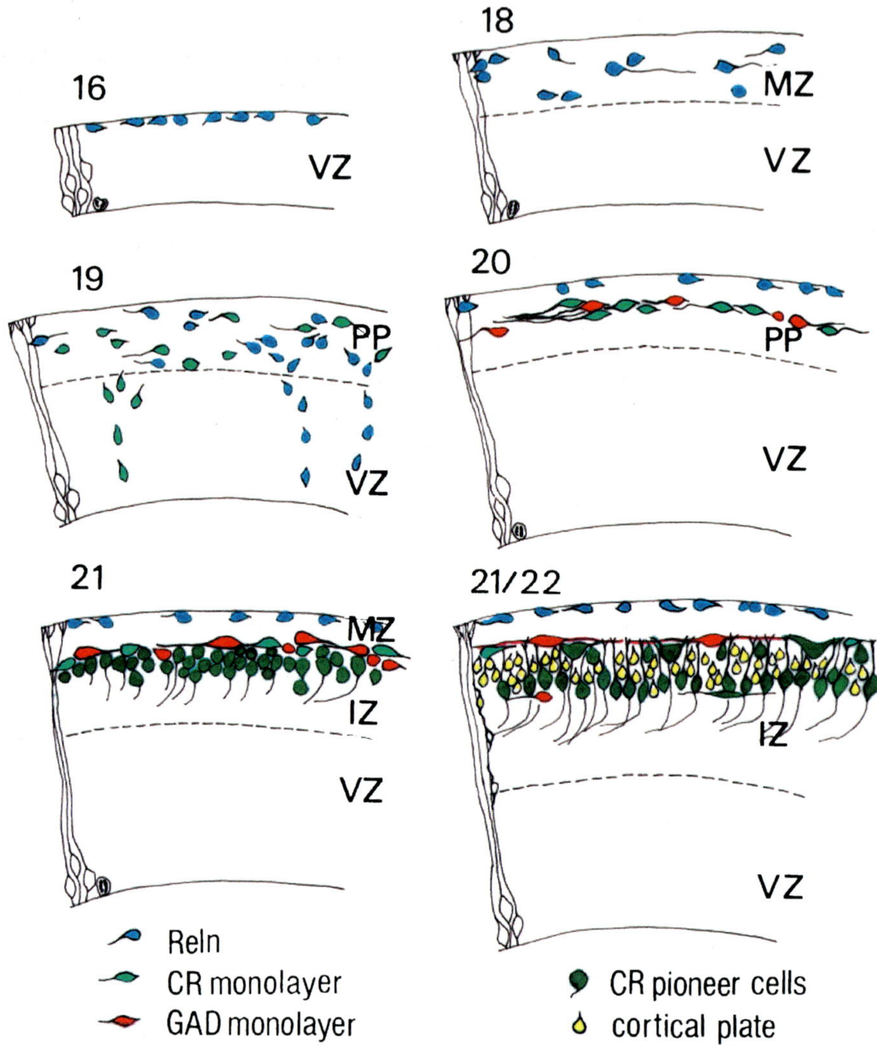

Fig. 9 Diagrammatic representation of the developmental events proposed here. In *blue*, Reelin; *light green*, early calretinin-positive cells in the monolayer; *red*, GAD-positive neurons of the monolayer; *dark green*, pioneer cells; *yellow*, cortical plate neurons. The first Reelin-positive neurons appear at CS 16 and increase in number from CS 17 to 19. The first calretinin-positive cells appear at CS 19 in what could now be called the preplate. Together with GAD-positive neurons, which appear at CS 20, they form the monolayer within the preplate. CR cells settle in the subpial compartment. At CS 21, the pioneer plate aggregates below the monolayer and sends the first corticofugal fibers. At CS 21/22, the pioneer plate is split apart into a minor superficial component and a large deep component, the subplate, through the first cohorts of the cortical plate. (From Meyer et al. 2000, with permission)

early human preplate. The initial moments of cortical development are crucial for understanding cortical malformations in humans. Since at present our most widely accepted experimental model is based on spontaneous or induced gene mutations in mice, the transition from preplate to cortical plate might need a closer inspection and reassessment in rodents.

5.3
Compartments of the Cortical Wall After the Appearance of the Cortical Plate

5.3.1
Ventricular/Subventricular Zone

During foetal development, the VZ diminishes in width concurrently with an increase of the SVZ. The SVZ appears first at 7 GW and increases in thickness from lateral to medial (Chan et al. 2002). It can be recognized by the presence of mitotic cells outside the VZ, and the smaller size and non- radial orientation of its cells. The dimensions and cellular organization of the primate SVZ are significantly different from the SVZ of lower mammals, in particular during the later stages of gestation when the VZ has mostly disappeared while the SVZ constitutes the main site of neuronal and glial proliferation. However, the SVZ is not uniformly distributed throughout the cortical wall. As noted by Zecevic et al. (2005), the appearance of the SVZ changes along the rostrocaudal axis as well as between the medial and lateral walls, and at any level, the medial SVZ is much narrower than the lateral SVZ, indicating considerable regional variability. In the monkey occipital lobe, the SVZ can be subdivided into an inner and an outer SVZ, separated by a fiber layer, and cells in the outer SVZ are arranged in a radial, "palisade-like" fashion (Smart et al. 2002). This distinct, radially orientated SVZ of the visual areas may be considered a primate-specific feature (Lukaszewicz et al. 2005). In our human material, the thickness of the SVZ was highest near the ganglionic eminences, where a "palisade-like" arrangement was not readily apparent. It is thus possible that the differentiation of the SVZ may be quite variable across the frontal, parietal, temporal, and occipital lobes, perhaps related to the specific lamination pattern and connectivity of the various architectonic areas. The regional variations of the SVZ may in part explain the differences in the definition of the SVZ in the work of Smart et al. (2002) and our own work (Meyer et al. 2002). On the other hand, the human primary visual cortex, Brodmann's area 17, is known as an extreme representative of a koniocortex, where the granular layers IV and II achieve their highest degree of differentiation (von Economo and Koskinas 1925). It is possible that the SVZ of the prospective visual cortex reflects the unique cytoarchitectonic features of the adult area 17.

Recent work in rodents has in fact established that precursor cells in the SVZ generate the neurons destined to the supragranular layers of the cortex, which would suggest a correlation between the variations in thickness of the SVZ and the granular or agranular phenotype of the overlying cytoarchitectonic area. Progenitor cells in the SVZ express a number of molecules that indicate a glutamatergic

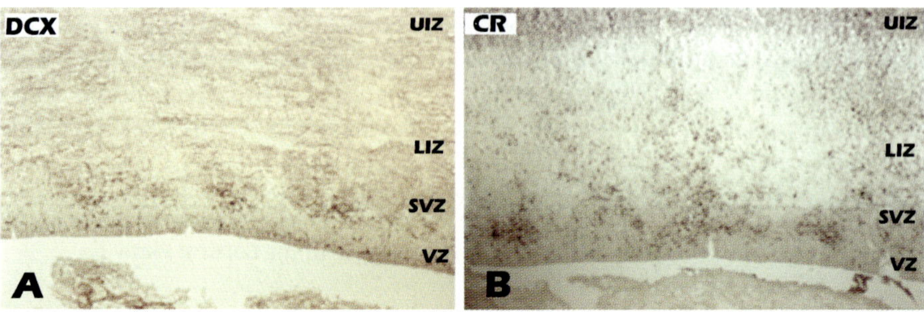

Fig. 10A, B Clusters of (**A**) DCX- and (**B**) calretinin (CR)-positive neurons in the SVZ at 14 GW. These presumptive interneurons display a horizontal orientation in the SVZ and then spread tangentially into the lower intermediate zone (*LIZ*). In the upper intermediate zone (*UIZ*), they can change into a more radial orientation

pyramidal phenotype (Tarabykin et al. 2001; Zimmer et al. 2004; Wu et al. 2005). Also in the human brain, the SVZ is the major proliferative compartment for neurons of the upper layers; it contains various cell types which contribute to both neurogenesis and gliogenesis (Zecevic et al. 2005). An important and unique aspect of the human SVZ is its potential to generate GABAergic interneurons. In rodents, apparently all interneurons are born in the ganglionic eminences of the ventral telencephalon (reviewed by Kriegstein and Noctor 2004), whereas in humans, a substantial proportion of GABAergic interneurons may originate in the cortical SVZ (Letinic et al. 2002; Rakic and Zecevic 2003). This possibility would indicate quantitative as well as qualitative differences in the cell composition and proliferative role of the SVZ among species.

In our studies on Doublecortin (DCX; Sect. 9.4.2), we observed at 14/16 GW clusters of postmitotic neurons in the SVZ expressing calretinin (Fig. 10B), a marker of cortical interneurons (Rogers 1992; Kubota et al. 1994; Hof et al. 2000), and DCX (Fig. 10A), which appeared to derive from the cortical SVZ (Meyer et al. 2002), supporting a pallial origin of subsets of interneurons.

The evolutionary expansion of the human cortex has been explained by an amplification of the founder pool of cortical precursors and changes in proliferation kinetics that increase the number of radial columnar units without changing cortical thickness (Rakic 1995). The increasing prominence of the SVZ during mammalian evolution may have contributed to the progression in the number of radial units characteristic of the primate lineage.

5.3.2
The Intermediate Zone

The intermediate zone (IZ) represents the fetal white matter and is a transit zone through which cortical neurons migrate radially and tangentially on their way to the cortical plate. When migration is completed, the IZ transforms into the adult white matter. In our material, we observed the first afferent fibers in the IZ at

8/9 GW, when the cortical plate is established in the lateral and dorsal neocortex (Fig. 2). Most of these fibers appeared to have their origin in the thalamus, but we cannot exclude that early fibers from the brain stem also travel through the IZ. As gestation proceeds, the IZ continuously increases in width and is particularly prominent in those cortical regions close to the ganglionic eminences and the internal capsule, whereas it is much narrower in the medial occipital lobe and in the cingulate cortex. The orientation and density of the fetal fiber tracts, visualized with calbindin (Fig. 11A) and calretinin (Fig. 11B), allow a subdivision into a lower intermediate zone and an upper intermediate zone.

At midgestation, the lower IZ contains numerous dividing cells which make it difficult to clearly delimit the IZ from the SVZ (Fig. 11C). The border between the IZ and the subplate can be recognized by a dense horizontal fiber plexus indicating the location of the external capsule (Kostovic 1990), and MRI imaging of fetal brains shows a sharp decline in signal intensity at the subplate/IZ border (Kostovic et al. 2002).

The IZ is a cell/fiber compartment that has attracted much less attention than others, and only more recent studies show that it serves as a migratory route for interneurons from the ganglionic eminence to the cortex (DeDiego et al. 1994; Tanaka et al. 2003; Metin et al. 2000; reviewed by Metin et al. 2006). A large variety of attractive/repulsive signals are involved in tangential migration (see Sect. 4.2), indicating that many cellular and molecular interactions that take place in the IZ are important for orchestrating tangential and radial migration pathways.

The white matter of the adult human cortex contains a population of "interstitial cells" which were thought to be remnants of the subplate (see Sect. 5.3.3.3). In our human fetal material at midgestation, we observed large, differentiated neurons in the IZ that strongly expressed Dab1 mRNA and protein (Fig. 23). This raises the question of whether the IZ contains a resident cell population that during development may serve as guideposts for migrating neurons. IZ cells may also survive into adult life, forming part of the interstitial cells of the deep white matter (Meyer et al. 1992).

5.3.3
The Subplate

Human cortical development is a slow process, and during several months the cortical plate is morphologically and functionally immature. During this period of immaturity, the subcortical fiber systems, destined to establish synaptic contacts with neurons in the prospective cortical layers, temporarily reside in the subplate and interact with transient neurons that form a "waiting compartment" at the base of the cortical plate. There is ample evidence from a variety of studies in rodents, carnivores, and monkeys that the connections between thalamus and cortex require the presence of subplate cells that mature early and participate in functional circuits (reviewed by Allendoerfer and Shatz 1994). Pioneer cells in the subplate emit the first corticofugal projections and are the target of the first thalamocortical fibers,

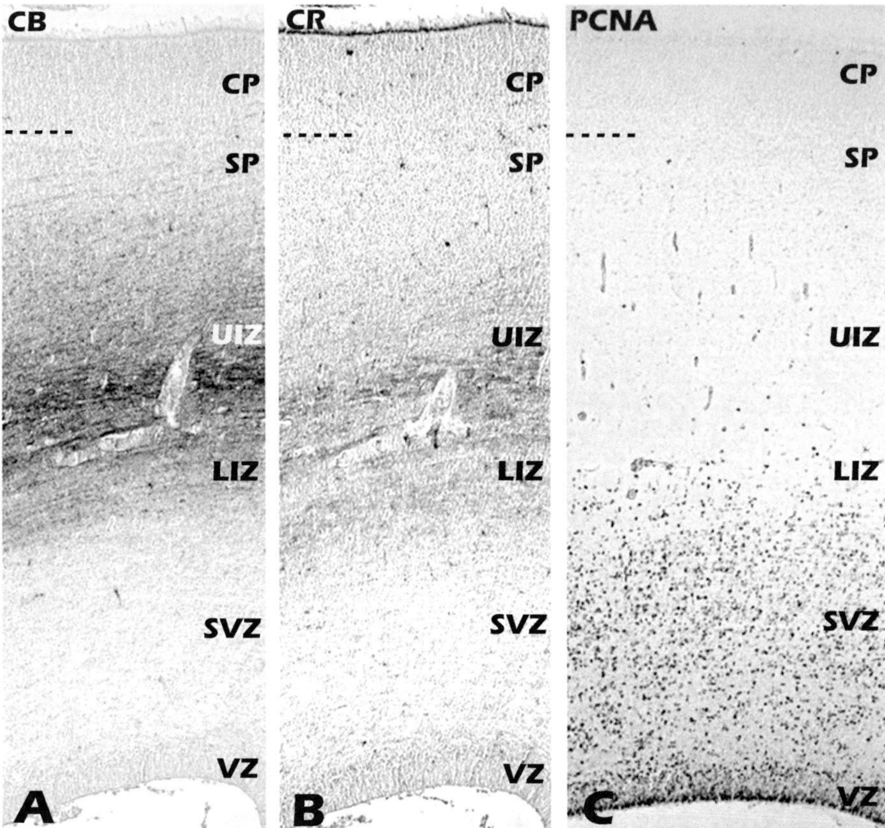

Fig. 11A–C The stratification of the cortical wall at 15 GW. **A** Calbindin and **B** calretinin mark the fibers of the intermediate zone, which can be subdivided into an upper (*UIZ*) and lower (*LIZ*) part. **C** The subventricular zone (*SVZ*) is highly developed at this stage and characterized by numerous mitotic figures, visualized with PCNA. Mitoses extend also into the *LIZ*, and the boundary between *SVZ* and *LIZ* is difficult to establish. At this age, the subplate (*SP*) is not yet fully developed. *CP* cortical plate

and the first synapses are established in the subplate (Molliver et al. 1973; Kostovic and Molliver 1974; De Carlos and O'Leary 1992; McConnell et al. 1989). In the kitten, transient NPY-positive subplate neurons, the so-called "axonal-loop cells", send axonal projections into the MZ (Wahle and Meyer 1987). Similar observations by Friauf et al. (1990) showed that these axonal processes participated in synaptic microcircuits. In the visual cortex, subplate neurons are involved in the formation of ocular dominance columns and orientation columns (Gosh and Shatz 1992; Kanold et al. 2003). The function of the subplate is thus crucial for the structural and functional development of the cortical plate.

In human and monkey brains, the arrival of afferent fibers in the subplate was assessed by using acetylcholinesterase (AChE) histochemistry (Kostovic 1986,

1990b; Kostovic and Goldman-Rakic 1983; Kostovic and Rakic 1984). From 20 to 24 GW thalamic and basal forebrain afferents accumulate in the "waiting compartment" of the subplate, before thalamocortical fibers penetrate the cortical plate at 24–28 GW, where they assume a columnar distribution. Once the cortical target layers have reached maturity, subcortical fibers leave the subplate and approach their final innervation target in the cortex. The subplate neurons are thought to disappear after having fulfilled their developmental role (see Allendoerfer and Shatz 1994).

The term subplate is often used in an ontogenetic sense, referring to early-born neurons that carry out important pioneer functions by establishing early connections and architectonic patterns. On the other hand, the term is also used as a positional criterion, referring to the cell and fiber compartment between the cortical plate and the intermediate zone. As we shall see below, the ambiguous use of the term and the generalized application of rodent and carnivore data to the human brain have led to theories that contradict anatomical realities.

5.3.3.1
A Timetable of Human Subplate Development

In primates, the subplate is the most outstanding compartment of the fetal cortical wall. The first architectonic and AChE studies on the human and monkey subplate revealed several distinct histogenetic stages (Kostovic and Rakic 1980, 1990). Most importantly, Kostovic and Rakic (1990) demonstrated that the human subplate increases in size after the emergence of the cortical plate. They described an initial stage, the "presubplate," which is populated by polymorphic neurons and recognizable as a narrow, cell-sparse tissue below the incipient cortical plate at 12 GW. This "presubplate" may correspond to our deep pioneer neurons in Fig. 8I, J. The next stage, from 12 to 15 GW, is considered as the stage of subplate formation, when the deep part of the cortical plate transforms into a cell- and fiber-rich "upper subplate," which is continuous with the cell-sparse "lower subplate." It is after 15 GW when the human subplate undergoes a dramatic increase in width and architectonic differentiation, reaching peak values by 30–34 GW. At its maximum development, the width of the subplate is four times greater than the width of the cortical plate. After 35 GW, the subplate decreases in size and disappears in the first postnatal month. An important aspect of the subplate is its variable extension in different cortical areas: it is highly developed in the somatosensory cortex, but rather small in the primary visual cortex. Furthermore, the resolution of the subplate begins in the depths of the sulci and occurs much later at the crowns of the gyri, suggesting a possible relationship between subplate dynamics and cortical gyration (Kostovic and Rakic 1990).

The prominence of the human subplate and its timetable of development established by Kostovic and Rakic (1990) have been confirmed in numerous studies with a variety of methods. Golgi studies emphasized the precocious development of dendritic and axonal processes of subplate neurons compared to cortical plate neurons (e.g. Mrzljak et al. 1988, 1990). Immunohistochemical markers such as MAP2 and GAP-43 showed early expression in the subplate compared to a later staining in the cortical plate (Honig et al. 1996). The calcium-binding proteins calbindin and calretinin are abundantly expressed in the human subplate (Ulfig 2002). Calretinin-labeling was particularly prominent during the 5th and 6th gestational months, whereas calbindin-positive neurons appeared at around 25 GW. Calbindin-expressing neurons were morphologically heterogeneous and concentrated in the upper third of the subplate. Parvalbumin appeared rather late, after the 30th GW. Cell counts revealed that from 13 to 20 GW the cell content in the subplate increased by a factor of 3.6; cell numbers still continued to increase between 22 and 35 GW, and finally decreased after GW 35 (Samuelsen et al. 2003). The dynamic changes in the subplate can also be detected by magnetic resonance imaging and may be useful for assessing the degree of maturity of the fetus (Kostovic et al. 2002).

For our own immunohistochemistry study of human subplate development we used an antibody against Tbr1, a transcription factor highly expressed by early-born neurons in the MZ, deep layer I and subplate (Hevner et al. 2001). We find that in the initial cortical plate, at 10–11 GW, virtually all neurons express Tbr1 (Fig. 12A). With increasing gestational age, Tbr1-negative neurons destined to successively more superficial layers migrate into the cortical plate, while the earlier-born Tbr1-positive neurons of layer VI and subplate settle at the base of the cortical plate and in the subplate. These findings are in accordance with the observation by Bayer and Altman (1990) that the earliest-born neurons, the subplate (or rather presubplate) neurons, form part of the first condensation of the cortical plate, and then progressively segregate from the cortical plate when later-born cohorts migrate beyond them.

Although the subplate zone considerably expands from 14 to 25 GW, the density of Tbr1-positive neurons remains high, indicating that new Tbr1-positive neurons are added to the previous ones. In fact, birthdating studies in monkeys showed that subplate neurons are generated in parallel with cortical plate neurons and continue to migrate into the subplate after the appearance of the cortical plate (Smart et al. 2002). Only the continuous addition of new cells can explain the extraordinary prominence of the primate subplate, and especially of the human subplate, in late gestation. Accordingly, the term subplate in primates should be used only following positional and functional criteria, because it does not necessarily imply an early birth.

5.3.3.2
Is the Subplate a Derivative of the Preplate?

One of the main tenets of cortical development holds that the preplate contains CR cells and subplate neurons, which are split into the MZ and the subplate by the arriving cortical plate (Marin Padilla 1978). Birthdating studies in carnivores (Luskin and Shatz 1985; Chun and Shatz 1989a; Jackson et al. 1993) and rodents (Bayer and Altman 1990; König et al. 1977; Wood et al. 1992) confirmed that the

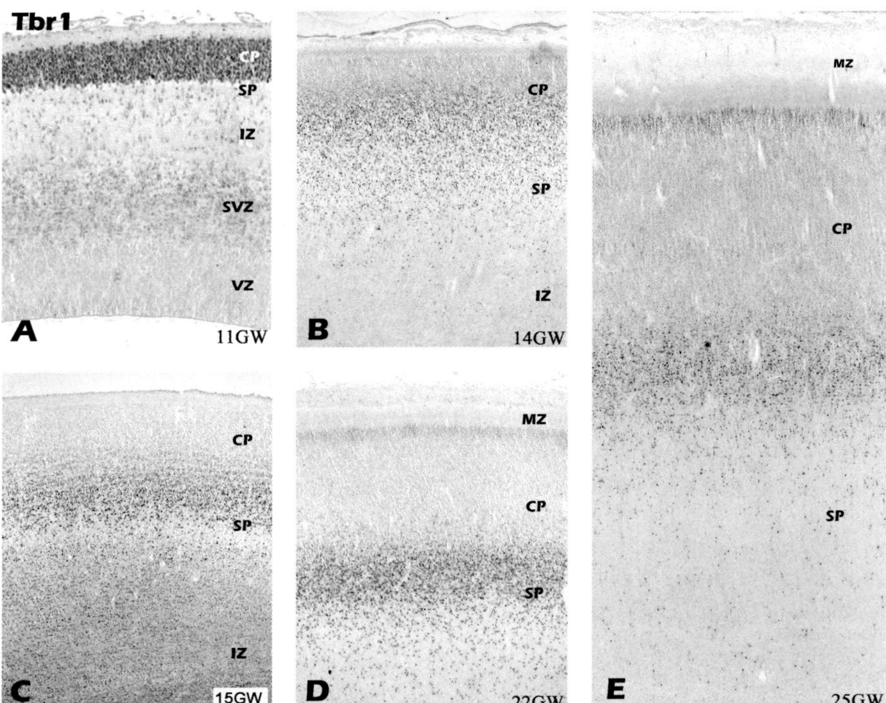

Fig. 12A–E Tbr1 expression and subplate development. **A** In the early stage of the cortical plate (*CP*) at 11 GW strong immunoreactivity is observed in almost all neurons that form the oldest layers of the cortex. The small subplate (*SP*) or presubplate is just below the CP. **B, C** At 14 and 15 GW the SP has increased in size; the upper part of the CP is Tbr1 negative. **D, E** During the following weeks both the CP and the SP increase in width. Tbr1-positive neurons are most numerous just below the CP, but also extend throughout the enlarged SP, which continues to grow. Since the cortex substantially increases in size during the period represented in this figure, it may be assumed that new subplate cells are added continuously to those present in the early stages

oldest cortical neurons are located in layer I and in the subplate deep to layer VI. As stated above, the subplate is crucial for corticogenesis by controlling and organizing important events in cortical-plate maturation and it is thus important to know its developmental origins.

An important aspect that has emerged in recent years is the heterogeneity of proliferation zones contributing to the cortex, as discussed in Sect. 4.2. GABAergic interneurons are generated in the ganglionic eminences and migrate tangentially from lateral to medial, whereas pyramidal cells are born in the cortical VZ and SVZ and migrate radially (Sect. 4.1). Most importantly, even the prototype of a preplate neuron, the CR cell, belongs to a heterogeneous cell family, and many CR cells reach the preplate by medial to lateral tangential migration from an extrapallial origin, the cortical hem (see Sect. 6; Meyer et al. 2002; Yoshida et al.

2006). In consequence, the preplate may be conceived as a site of convergence of multiple tangential and radial migration streams, rather than as a succession of radial migrations from a single cortical proliferative zone. At present, we do not know which cell components of the preplate have an extrapallial origin, since some pioneer neurons may also be generated outside the pallial VZ (Morante-Oria et al. 2003). It is thus very possible that the subplate may also derive from multiple generation sites. Interestingly, Zecevic and Milosevic (1997) described the presence of GABA-immunoreactive neurons in the subplate of 7–13-GW-old fetuses, which suggests a contribution of the ganglionic eminence to the subplate also.

As a first step toward a better understanding of the complexity of the subplate, it would be necessary to recognize the distinct stages of the primate subplate—presubplate, developing, and mature subplate—in nonprimate species also. Furthermore, it would be interesting to know whether rodents and carnivores have all the cell components of the primate cortex and undergo the same developmental steps, in order to assess whether the non-primate brain can really provide an experimental model for the human subplate.

At this point, the findings in monkey and human (Smart et al. 2002; Meyer et al. 2000 2002a) indicate that the current model of a preplate split into CR cells and subplate neurons is not in accordance with the anatomical facts in primates.

5.3.3.3
The Subplate as the Source of Neurons in the Adult White Matter

In keeping with its transient role, the subplate disappears at the end of corticogenesis (Kostovic and Rakic 1980, 1990; Allendorfer and Shatz 1994). In kittens, early-generated subplate neurons disappear during early postnatal life (Luskin and Shatz 1985; Chun and Shatz 1989a, b). Also in kittens, early maturing GABAergic and NPY-positive cell populations in the subplate degenerate and die in the first postnatal days (Wahle and Meyer 1987; Wahle et al. 1987). However, important species differences seem to exist. In the rat (Valverde et al. 1995), cell death in the subplate does not appear to be significant and layer VI might represent the homologue of the primate subplate. According to Robertson et al. (2000), cells of the rat subplate survive into adult life, although perhaps at a reduced functional state. A more differentiated view of the rodent subplate is offered by recent findings on the expression of a Tbrain-1 (Tbr1)-driven transgene in the deep cortical plate and subplate (Kolk et al. 2005). Transgene-expressing cells were born very early in corticogenesis, had complex dendritic trees during cortical plate formation, and thereafter underwent progressive simplification as they survived into adulthood as excitatory resident cells of deep layer VI and white matter.

The idea that remnants of the subplate persist as interstitial neurons of the adult white matter in human and monkey cortex was first proposed by Kostovic and Molliver (1974) and Kostovic and Rakic (1980, 1990). In fact, the human cortical white matter contains large numbers of neurons (Fig. 13), many of which have the morphology of pyramidal neurons, displaying apical and basal dendrites cov-

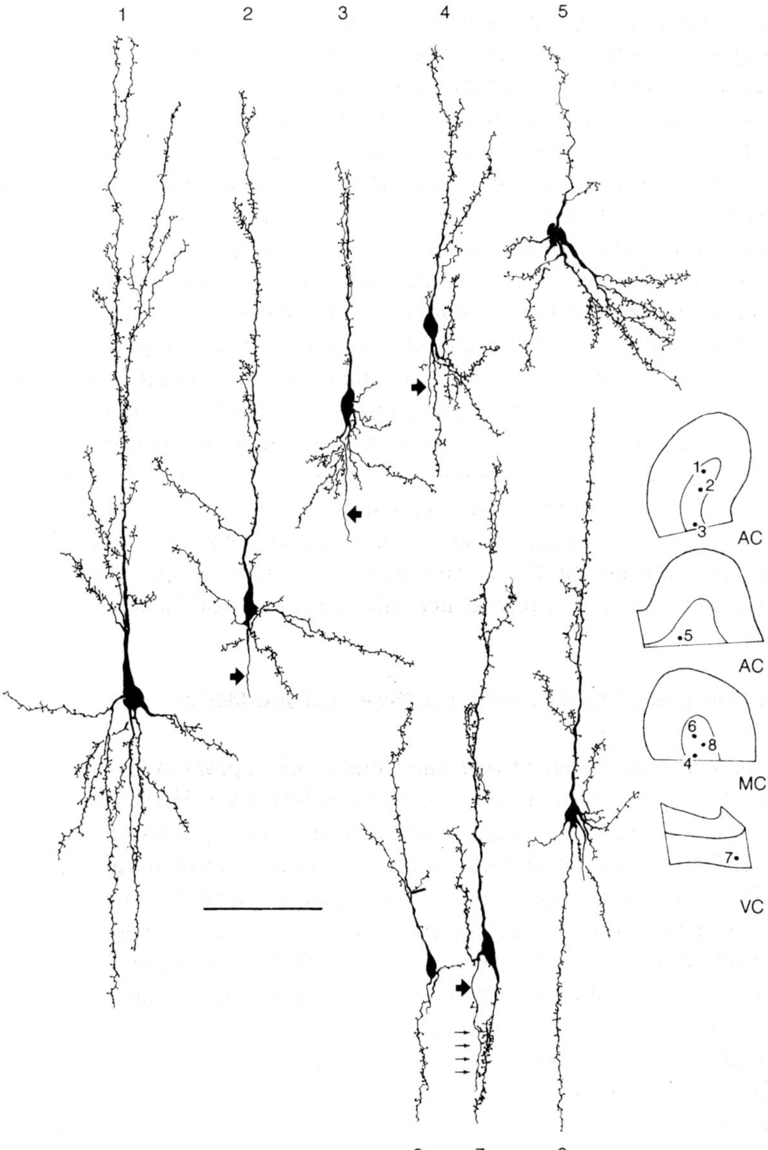

Fig. 13 Interstitial neurons of the adult white matter, stained with the Golgi method. These cells often display pyramidal-like morphologies with a bipolar, radial orientation of their apical and basal dendrites, covered with dendritic spines. They do not resemble the early-born neurons of the presubplate, and we propose that they represent late-arriving neurons of the subplate and intermediate zone. *AC, MC, VC,* auditory, motor, and visual cortex, respectively. (From Meyer et al. 1992, with permission)

ered with dendritic spines, although their polarity may be modified, particularly in the deep white matter (Meyer et al. 1992). Other interstitial neurons express neurochemical markers characteristic of nonpyramidal neurons, such as GABA, NADPH-diaphorase, or neuropeptide Y (Chun and Shatz 1989b; Meyer et al. 1992). Interstitial neurons are more numerous in the superficial white matter at the level of the former subplate, but substantial numbers also reside scattered among the deep fiber fascicles of the former IZ, and we suggested above that the IZ also contributes to the adult interstitial neurons. If we consider that the primate subplate is a huge cell and fiber compartment that still grows until the third trimester of gestation, it is reasonable to assume that not all subplate neurons die but instead survive as interstitial neurons that become integrated into permanent circuits. By contrast, if we choose to use the term subplate for the earliest-born neurons of the cortex, we can only state that due to the dimensions of the adult human white matter it is highly unlikely, or entirely impossible, that the interstitial cells are a relict of the presubplate or the deep pioneer cells.

Again, it is necessary to emphasize the clear species differences, because white matter neurons are prominent in the human brain, but sparse in rodents and carnivores (Meyer et al. 1991). Taking together the many disparate observations in rodents, carnivores, and primates, we arrive at the conclusion that the subplate—defined as the cell and fiber compartment just below the cortical plate—is composed of a large variety of cell populations born at different moments of corticogenesis, some of which are transient whereas others survive and maintain functional circuits into adulthood.

Why are the interstitial neurons of the adult white matter important? Studies of the brains of psychiatric patients suggested alterations in the density of interstitial neurons in mental diseases, in particular in schizophrenia (e.g., Akbarian et al. 1996; Anderson et al. 1996; Eastwood and Harrison 2005), although these alterations could not be confirmed by other authors (Beasley et al. 2002). Altered gene expression was also reported in white matter neurons in bipolar and depressed patients (Molnar et al. 2003). Based on the postulated ancient origin of the subplate, it has been proposed that the defect of white matter neurons might reflect a developmental defect that occurred early in development. It is therefore important to clarify the definition of the human subplate and to distinguish the varieties of white matter neurons, as well as their origins and time of birth. For a correct interpretation of abnormal numbers and distributions of interstitial neurons in pathological brains, it is also important to take into account the increasing size and complexity of the white matter compartment during evolution, as well as the possibility that generation of interstitial cells can be a late event in corticogenesis.

5.3.4
The Cortical Plate

The cortical plate is the final destination of migrating neurons. Together with the MZ, it forms the adult six-layered neocortex. The main anatomical aspects of the developing cortical plate have been mentioned in Sect. 3.3. The arrival of excitatory pyramidal cells and inhibitory interneurons and their coordinated positioning within the correct isochronic layers are complex events as yet poorly understood. Human cortical architecture can be disrupted in the most diverse ways, giving rise to an immense variety of malformations which have been classified into various categories in textbooks and reviews on neuropathology (e.g., Friede 1989; Barkovich et al. 2001; Aicardi 1991; Clark 2004; Francis et al. 2006). Since new gene mutations are continuously being discovered, and the analysis of cortical tissue obtained from genetically identified human brain malformations will become more and more available, in the future we will arrive at a more differentiated view of human cortical development.

The classical descriptions of the human cortex (Brodmann 1909, von Economo and Koskinas 1925) emphasized the differences between cytoarchitectonic areas, which are often functionally distinct and establish characteristic sets of connections. Since in the rodent brain the morphological heterogeneity of cytoarchitectonic areas is much less pronounced and quite difficult to recognize, the use of the mouse as a genetic model has led to a rather simplified view of the human cortex. Importantly, the aspects of architectonic diversity have been substituted by a broader concept of regionalization imposed by the expression of region-specific molecules, which have been extensively analyzed in rodents. Examples are Pax6, Tbr1, Dlx1,Dlx 2, Emx1, Emx2, the Wnt family, FGF8 (Bulfone et al. 1993; Bishop et al. 2000, 2002; Mallamaci et al. 1998, 2000; Gulisano et al. 1996; Lee et al. 2000; Muzio et al. 2002; Porteus et al. 1994; Price 1993; Simeone et al. 1992; Stoykova et al. 1996, 2000; Talamillo et al. 2003; Theil et al. 2000; Yoshida et al. 1997; Garel et al. 2003; Fukuchi-Shigomori and Grove 2001, 2003; Furuta et al. 1997; Cecchi and Boncinelli 2000).These molecules are thought to interact, to provide positional information, or to regulate regional growth. In a gene regulatory cascade, they may lead to the patterned differentiation of the cortical fields and finally help to construct the cortical map (Grove and Fukuchi-Shimogori 2003; Shimogori et al. 2004; Rash and Grove 2006; Hamasaki et al. 2004; Huffman et al. 2004; Sansom et al. 2005; Mallamaci and Stoykova 2006; O'Leary and Nakagawa 2002), in part through mechanisms intrinsic to the neocortex (Miyashita-Lin et al. 1999).

Thus far, very little is known on the expression of region-specific genes in the human or primate cortex (e.g., Donoghue and Rakic 1999; Abu-Kahlil et al. 2004; Lindsay et al. 2005), but undoubtedly more information will soon be available and hopefully lead to a reconciliation of the classical cytoarchitectonic fields with molecular regionalization. Furthermore, important cortical functions in human brain are lateralized, and interhemispheric molecular differences are recognizable at early embryonic stages and reflected by early, marked transcriptional asymmetries. Several genes including LMO4 are consistently more highly expressed in the right perisylvian human cerebral cortex than in the left, suggesting that human left-right specialization reflects asymmetric cortical development at early stages (Sun et al. 2005).

The concept that the multiple architectonic fields of the primate, and probably also of the human neocortex, are under the control of area-specific regulatory genes is emerging from studies showing cell-cycle differences among specific areas (Rakic 1988,1995; Kornack and Rakic 1998; Dehay et al. 1993; Kennedy and Dehay 1993). For instance, differential modulation of cell-cycle-related mechanisms leads to the emergence of primate areas 17 and 18, which display striking differences in cytoarchitectonics and neuron number. In area 17, precursors of supragranular neurons exhibit a shorter cell cycle duration, a reduced G1 phase, and a higher rate of cell cycle reentry than area 18 precursors. Furthermore, area 17 and area 18 precursors show contrasting and specific levels of expression of cyclin E and p27Kip1 (Lukaszewicz et al. 2005).

5.3.5
The Marginal Zone

The predecessor of cortical layer I or molecular layer is known as the marginal zone (MZ) and is located underneath the pia. The endfeet of astrocytes form the external limiting membrane, which is in contact with the meninges. The marginal zone appears very early in development, and in the rat is also the first layer where axo-somatic synaptic contacts are established on Cajal–Retzius cells (König et al. 1975). During development, the most striking cell population in the MZ are the CR cells, which will be discussed extensively in Sect. 7.3.1. CR cells are transient, even though a few remnants are occasionally found in the adult. In addition, in rodents there are other transient components of the MZ, such as the superficial pioneer cells described in Sect. 5.2, and, most prominently in humans, the cells that form the subpial granular layer. Since the SGL has unique features in the human brain, we will discuss them here in detail. Fibers in the definitive molecular layer have many diverse origins: the sensory, midline and intralaminar thalamic nuclei, the noradrenergic and serotoninergic brain stem centers, cholinergic cell groups of the basal forebrain, and ipsi- and contralateral cortical areas (reviewed by Vogt 1991), although it is not known at what moment of development they arrive. The apical dendritic tufts of pyramidal cells are important structural elements of the underlying layers (Marin Padilla and Marin-Padilla 1982). Layer I of the adult cortex is remarkably cell poor and populated only by small GABAergic interneurons (Gabbott and Somogyi 1986) and glia.

5.3.6
The Subpial Granular Layer

The SGL is an enigmatic transient cell layer in the uppermost tier of the MZ just beneath the pial surface. As yet, the SGL has been described only in the human cortex and it is not known whether it exists in non-primate species also. We owe the first description of the SGL to Ranke (1909) who termed it the "superficial granular layer" (superfizielle Körnerschicht). Brun (1965) coined the now generally

at the level of the basal forebrain of a 27-GW-old fetus. A dense stream of migratory cells leads from the rostral edge of the lateral ventricle to the olfactory bulb, and probably represents the rostral migratory stream (RMS), the migration route of the interneurons of the olfactory bulb (Menezes et al. 1995). In addition, a second, slightly more caudal and more dispersed stream extends from the SVZ of the ganglionic eminence to the basal surface of the anterior perforated substance (APS), often guided by large blood vessels. This stream is continuous with a thick SGL that in turn leads to the MZ of the *limen insulae*, the anterior border of the insular cortex, confirming the earlier descriptions by Brun (1956) and Gadisseux et al. (1992). These anatomical observations lend support to a possible migratory pathway of cortical interneurons from the most medial edge of the ganglionic eminence to the neocortex via the SGL. The close spatial relationship between CR cells and SGL cells suggest a potential interaction between the two cell populations. CR cells express the extracellular matrix protein Reelin, which may provide guidance cues for the migrating interneurons to reach their final areal position.

This hypothesis has not been tested experimentally, because the SGL is thought to be specific to primates or humans and thus difficult to access. However, Ranke (1910) mentioned that the SGL is also present in carnivore brains. Even in the rodent cortex a rudimentary SGL has been recognized (Meyer et al. 1998, 1999), and interestingly enough, many superficial granule neurons in the embryonic rat cortex at E 16 express GAD (Fig. 5F). It may be hoped that future experimental approaches will shed light on the role of the SGL.

6
The Cortical Hem: Signaling Center and Birthplace of CR Cells

The human hippocampal anlage is an annular structure along the midline of the hemisphere, and its border with the choroid plexus epithelium forms the limbus or medial edge of the cerebral cortex, according to the classical definition by Paul Broca (1878). This concept of the cortical limbus or hem has been further developed by Grove (Grove et al. 1998; Grove and Tole 1999; Ragsdale and Grove 2001) who re-defined the medial edge of the cortex as an important signaling center for the adjacent cortex during early stages of development. In the mouse, the cortical hem is defined by the graded expression of a variety of genes, most prominently of the Wnt and BMP families. The cortical hem is adjacent to the roof plate and necessary for the development of the hippocampus (Lee et al. 2000) and the regionalization of the dorsal cortex (Ragsdale and Grove 2001). It was found that cell migrations from the roof plate might influence the size of the future cortical territory (Monuki et al. 2001; Monuki and Walsh 2001). We observed that in embryonic mice at E 11/12 the cortical hem gave rise to large numbers of p73/Reelin-positive cells that migrated tangentially into the neocortical preplate (Meyer et al. 2002a) Since in embryonic p73$^{-/-}$ mutant mice these cells were missing, and the p73$^{-/-}$ cortex is devoid of CR cells, we proposed that the cortical hem is the origin of the most substantial population of neocortical CR cells (Meyer et al. 2002a). Recent genetic ablations of the cortical hem confirmed that CR cells are born in the hem; they were absent in hem-ablated mice (Yoshida et al. 2006). As in the case of p73-deficient mice, cortical layering was not disrupted in these mutants, bringing into question the role of CR cells in the lamination process.

The human cortical hem is much more prominent and differentiated than the mouse hem, and the numbers of p73-positive CR cells originating from the hem are also much larger (Abraham et al. 2004). This observation raises several questions, thus far not fully addressed. Do the CR cells participate in the signaling activities attributed to the cortical hem, and are they involved in the patterning roles of the hem? Are CR cells identical to the cells observed by Monuki et al. (2001) that may influence cortical size? Is the tumor protein p73 (Kaghad et al. 1997) involved in the signaling cascades of the cortical hem? In fact, our observations that the cortical hem of embryonic mice expresses the transactivation-competent TAp73 isoform, whereas CR cells express the non-transactivating truncated DeltaNp73 isoform (Meyer et al. 2004), point to a complex role of p73 in cortical hem signaling. TAp73 is activated by the transcription factor E2F1 (Irwin et al. 2000). The similarities in the phenotypes of E2F1-knockout mice and p73-heterozygous mice, with both showing an alteration of caudal cortical areas, are in keeping with a distinct role of an E2F1/TAp73-signaling pathway in cortical regionalization (Meyer et al. 2004).

The human cortical hem is inconspicuous in early preplate stages (Fig. 17A, B) and reaches its maximum development around the time of most intense CR cell production, at 8/9 GW (Fig. 17C–E). A comparison of Fig. 17C and D shows that CR cells express only p73 in the hem, and begin to coexpress p73 and Reelin when they reach the prospective hippocampal fissure and then the MZ of the medial cortex. The hem regresses and stops producing CR cells when the hippocampus begins to develop. After the appearance of the dorsal hippocampal primordium, at 9 GW, there are no longer p73-positive cells in the cortical hem (Fig. 17F), which instead is occupied by pioneer fibers of the fimbria (Abraham et al. 2004). The generation of CR cells in the cortical hem thus takes place within a tightly controlled time window which may be regulated by specific events in the adjacent cortical plate and hippocampal neuroepithelium. Our timetable and the site of p73 expression match the expression of the hem-specific genes Wnt2b and Bmp7 reported by Abu-Kahlil et al. (2004) in the human brain, indicating highly conserved functions of the cortical hem in humans and mice. Certainly, the hem has so far not received the attention it deserves, and further studies may provide surprising insights into cortical patterning and the role of CR cells in the limbic telencephalon.

7
The Reelin–Dab1 Signaling Pathway

The Reelin–Dab1 signaling pathway plays a key role in cortical development. Its main components have been identified genetically and biochemically, and its cel-

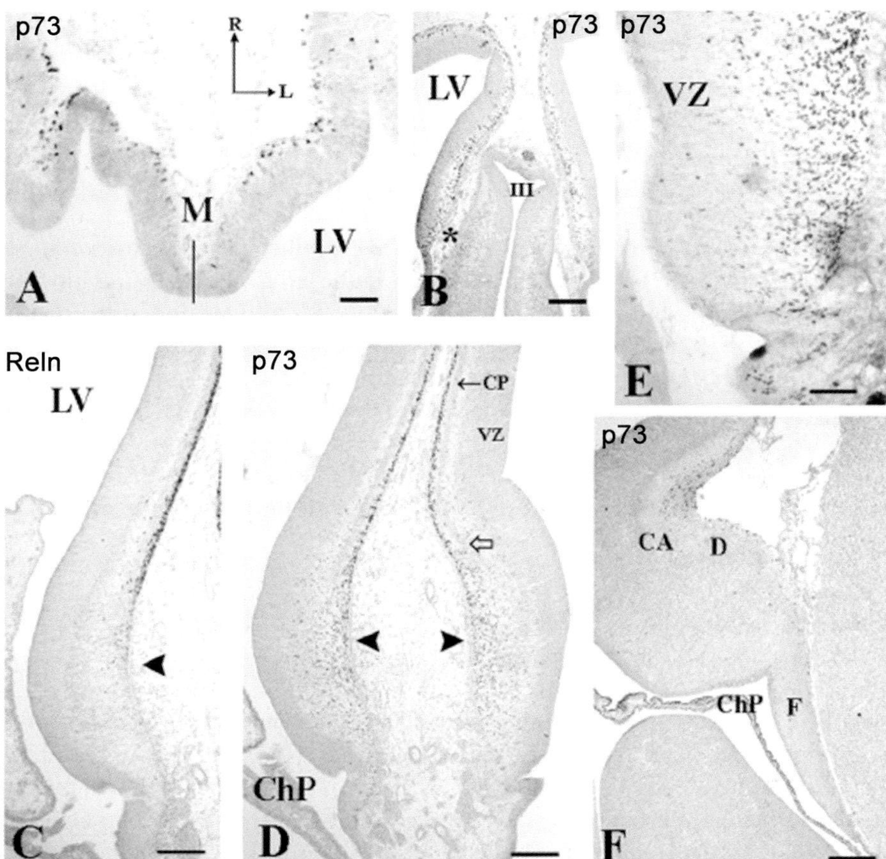

Fig. 17A–F The cortical hem as the origin of CR cells. **A** At 6 GW (CS 17). Horizontal section through the rostro-medial hemispheric wall. Expression of the CR cell marker p73 is still weak near the midline (*M*), but high in the marginal zone. *R*, rostral; *L*, lateral. **B** Goronal section at 7 GW (CS 20) showing p73-positive cells in the medial cerebral wall. The interface of the cortical hem and the choroid anlage (*asterisk*) is marked by numerous p73-positive cells in the deep VZ. **C–E**: At 8 GW (CS 22). **C** Reelin (*Reln*) and **D** p73 in panoramic views of the cortical hem (*arrowheads*) and the adjacent cortex prior to the overt appearance of the hippocampus, the moment of highest production of CR cells. *Arrow* in **D** indicates the cortical plate (*CP*), *open arrow* the ventral edge of the CP. **E** High magnification of the cortical hem showing large numbers of p73-positive cells in the VZ and MZ of the cortical hem. **F** At 14 GW. After the appearance of the hippocampus and fimbria (*F*), the cortical hem undergoes regression and does not produce any more CR cells. *CA*, cornu ammonis; *D*, dentate anlage; *ChP*, choroid plexus. *Scale bars*: **A** 50 µm; **B** 175 µm; **C, D** 100 µm; **E** 50 µm; **F** 200 µm. (Adapted from Meyer et al. 2002, with permission)

lular substrates are well characterized (reviewed by Tissir and Goffinet 2003). The signaling cascade is initiated by the extracellular matrix protein Reelin, which is secreted by CR cells of the marginal zone. The Reelin signal acts through the extracellular milieu on target neurons in the cortical plate, regulating their positioning in the correct layer. The response to Reelin requires the combined presence of two lipoprotein receptors, namely VLDLR (very low density lipoprotein receptor) and ApoER2 (apolipoprotein E receptor type-2; Tromsdorff et al. 1999), and of the intracellular adapter protein Disabled 1 (Dab1; Howell et al. 1997; Sheldon et al. 1997; Ware et al. 1997). Reelin binds directly to the extracellular ligand binding domains of VLDLR and ApoER2, whereas Dab1 docks to the cytoplasmic region of the receptors containing an NPxY motif. The Reelin signal is transduced by tyrosine phosphorylation of Dab1. Downstream effects of the signaling pathway perhaps act by influencing the organization of the cytoskeleton (Hiesberger et al. 1999). During corticogenesis, VLDLR, ApoER2, and Dab1 are expressed in neurons of the cortical plate while the Reelin signal is confined to CR cells in the MZ, and this spatial segregation of Reelin-producing and Reelin-target cells may be relevant to the mechanisms of action of Reelin.

7.1
Reelin Gene and Protein

The *Reelin* gene is about 450 kb long and maps to mouse chromosome 5 and human 7q22 (De Silva et al. 1997; Royaux et al. 1997). The Reelin protein is a large glycoprotein (predicted molecular mass 388 kDa) that is secreted into the extracellular matrix. It contains 3,461 amino acids that form a distinct N-terminal region, followed by eight internal repeats of 350–390 amino acids, and a short, positively charged C-terminal region (D'Arcangelo et al. 1995; Lambert de Rouvroit and Goffinet 1998a). The full-length protein is rarely detected in brain extracts and body fluids, where N-terminal fragments of 320 and 180 kDa and C-terminal fragments of 240 and 100 kDa are the dominating forms (Tissir and Goffinet 2003). Reelin is cleaved in vivo; a metalloproteinase is involved (Lambert de Rouvroit et al. 1999). The central region of Reelin comprising at least four repeats is essential for receptor binding and triggering of Dab1 phosphorylation (Jossin 2004). Secretion occurs through a constitutive pathway, independently from storage in synaptic vesicles and neurotransmitter regulation (Lacor et al. 2000), and requires the C-terminal region (D'Arcangelo et al. 1997). In *reeler Orleans* (*Reln^{Orl}*) mutant mice, this latter region is disrupted and Reelin is expressed but not secreted (de Bergeyck et al. 1997), and the animals display the same phenotype as *Reelin* null mutants. The N-terminus of Reelin is not directly involved in receptor binding (Hiesberger et al. 1999), but may be needed for Reelin aggregation, which is inhibited by the CR50 antibody (Utsunomiya-Tate et al. 2000), directed against an epitope in the N-terminal region (Ogawa et al. 1995). Reelin is widely distributed throughout the developing and adult brain and a large variety of other organs (Ikeda and Terashima 1997).

7.2
The Effects of Reelin Deficiency in Mice and Humans

The Reelin-deficient *reeler* mouse is the classic model of a gene defect that disturbs fundamental processes in brain development leading to a characteristic abnormality of brain architecture. During several decades the *reeler* mouse was studied extensively as a unique model of an inverted cortical migration gradient (reviewed by Lambert de Rouvroit and Goffinet 1998), although the causative gene defect was not known until the cloning of the *reelin* gene was reported by D'Arcangelo et al. in 1995. Several alleles have been described in mice and rats, and all show the same behavioral and anatomical phenotype. Ataxic gait, tremor, and dystonia are associated with a severe hypoplasia of the cerebellum; they are accompanied by alterations of the architecture of other laminated brain structures, in particular of the cerebral cortex (Caviness and Sidman 1973; Goffinet 1984). The *reeler* phenotype is also produced by mutations of the *Disabled 1* (*Dab1*) gene, by double mutations of the two high-affinity Reelin receptor genes *VLDLR* and *ApoER2*, and by other mutations in which the Reelin–Dab1 signaling pathway is disrupted or tyrosine phosphorylation of Dab1 abolished (Tissir and Goffinet 2003; Yossin 2004). All these mutations are described as causing a *reeler*-like phenotype.

In the *reeler* cortex, defective Reelin signaling apparently does not affect the initial formation of the preplate. However, in the subsequent step, the cortical plate neurons do not aggregate within the preplate and its associated extracellular matrix, and the preplate remains as an undivided superficial layer, termed the "superplate" (Fig. 18G; Caviness 1982; Sheppard and Pearlman 1997). Nonetheless, CR cells of $Reln^{Orl}$ mice can be visualized by immunostaining, using N-terminal antibodies, and they occupy their normal position below the pial surface (Fig. 18D).

Later-born cortical plate neurons are unable to migrate past earlier-born cohorts and instead settle below them, so that migration proceeds following an outside-in gradient. In consequence, the cortex is grossly inverted and a marginal zone is absent. The oldest neurons lie beneath the pia and the youngest neurons are adjacent to the white matter. On the whole, lamination of the reeler cortex is rudimentary, and neurons of the $Reln^{Orl}$ mice born at a given stage are more widely dispersed in the radial dimension than in wild-type mice. Many cortical plate neurons are obliquely oriented and less densely packed than in wild-type animals. The abnormal lamination is paralleled by an abnormal fiber distribution. Afferent fiber fascicles traverse the cortex obliquely until they arrive at the superplate, and then arch to reach their target in the cortical plate (Fig. 18G; Caviness 1976). Despite the severe architectonic malformation, the specificity of connectivity is remarkably conserved. This is the case of callosal, thalamocortical, and corticospinal fiber pathways (Molnar et al. 1998; Steindler and Cowell 1976; Terashima et al. 1983, 1985, 1987). The migration of interneurons to the superplate and their subsequent descent into the cortical plate is not affected by Reelin deficiency, and the coordination between laminar fates of interneurons and pyramidal cells is preserved

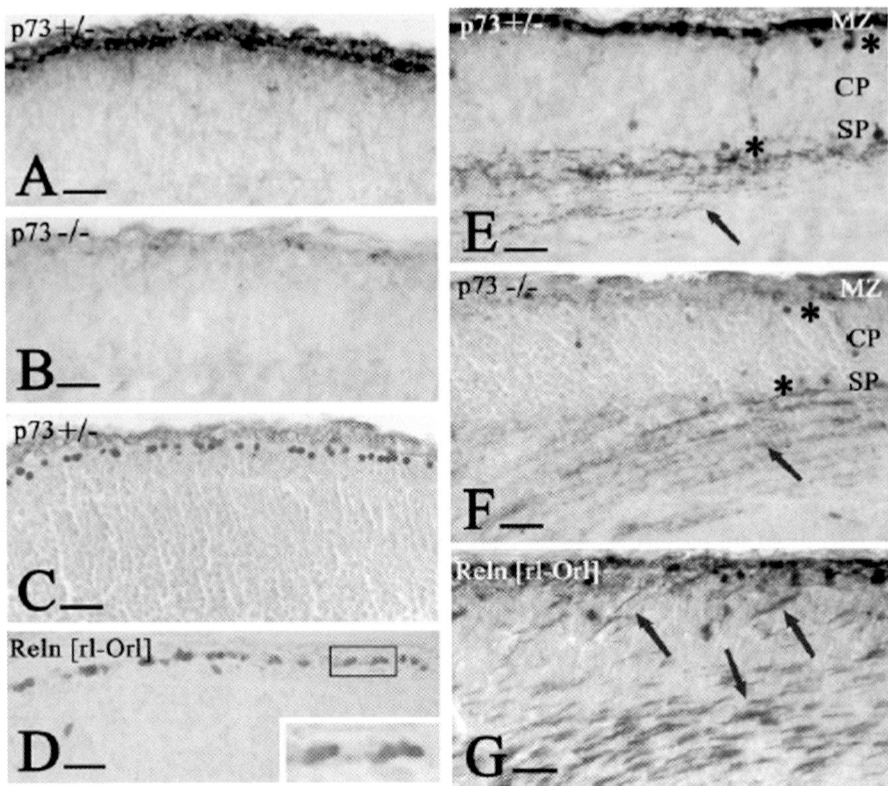

Fig. 18A–G Early development of the cortical plate in mouse embryos in the absence of CR cells. **A** In p73$^{+/-}$ mice, embryonic day (E) 15. **B** E 15 p73$^{-/-}$ mice, Reelin immunostaining. CR cells form a dense cell layer below the pial surface in **A**, but are absent in **B**, where only few faintly Reelin-positive cells are in the MZ. **C** E 15 p73$^{+/-}$ mice, showing high expression of p73 in CR cells. **E** More p73$^{+/-}$ mice. **F** In p73$^{-/-}$ mice, calretinin. CR cells are calretinin positive and are absent in the p73$^{-/-}$ cortex. The *asterisks* mark calretinin-positive pioneer neurons above and below the cortical plate (*CP*), and *arrows* point to thalamocortical fibers terminating in the subplate (*SP*); both elements are unchanged in the mutant brain, indicating that preplate splitting is not affected by the absence of CR cells. **D, G** *Reln*$^{rl-Orl}$ mice, in which Reelin and p73 are coexpressed in CR cells (**D**). In E, calretinin stains superplate elements and aberrant thalamocortical fibers (*arrows*) traversing the abnormal CP. Compare with **F**, in which CP and thalamocortical fibers have normal positions. *Scale bars*: 25 μm. (From Meyer et al. 2004, with permission)

(Hevner et al. 2004). Interneurons extend abnormal, hypertrophic processes, although maintaining a correct neurochemical profile (Yabut et al. 2005).

The radial glia morphology is disturbed: the number of radial glial cells with long radial processes is significantly decreased, pial endfeet are less arborized and the external limiting membrane formed by these endfeet is partially disrupted (Hunter-Schaedle 1997; Derer 1979; Pinto-Lord et al. 1982; Hartfuss et al. 2003).

However, neurogenesis is not affected, and the number of neurons is identical in the wild-type and reeler mutant cortex (Hartfuss et al. 2003).

Interestingly, the severe malformations of the reeler brain are not accompanied by epilepsy. The cortical malformation is also surprisingly well tolerated by the animals, which have no apparent learning or memory defect, with the cerebellar syndrome being the most evident behavioral disorder. This is in stark contrast to mutations of the human *reelin* gene, which are characterized by severe mental retardation and epilepsy. In humans, Reelin-deficiency causes an autosomal recessive form of lissencephaly associated with cerebellar hypoplasia (LCH). Two independent mutations of the human gene encoding Reelin have been identified (Hong et al. 2000). The affected patients found in two consanguineous pedigrees showed moderate lissencephaly, profound cerebellar hypoplasia, and abnormalities of hippocampus and brainstem. Cognitive development was severely delayed with little or no language, and the patients had no ability to stand or sit unsupported. All children had generalized epilepsies. At present, the phenotype is known only from medical imaging. Post-mortem studies of the architectonics of human Reelin-deficient brains are not available, and it is thus not known whether cortical lamination is affected in the same way as in the *reeler* mouse. In any case, the discrepancy between the impact of Reelin deficiency on human and mouse brain development indicates that Reelin plays more significant roles in the human than in the rodent brain, which is in keeping with the hypothesis that Reelin has been a driving factor in evolution (Bar et al. 2000; Tissir et al. 2002).

It has been suggested that the heterozygote reeler mouse (haploinsufficiency for the *reelin* gene) shares several neurochemical and behavioral abnormalities with schizophrenia (Pappas et al. 2001). The downregulation of Reelin synthesis has been proposed to lead to a decreased rate of secretion of the protein into the extracellular space and to an inhibition of maturation and plasticity of dendritic spines. In the heterozygous reeler mouse, there is also a decrease of Reelin- and GAD-positive neurons and an increase in neuronal packing density, reminiscent of findings in the brains of schizophrenic patients (Liu et al. 2001; Guidotti et al. 2000).

The *reeler* phenotype has lent support to the notion that the preplate is split into a MZ and a subplate in a Reelin-dependent fashion. In the absence of Reelin in CR cells, the preplate fails to split giving rise to the "superplate." It is important to note here that the cell composition of the "superplate" has so far not been studied in detail, and the position and fate of the superficial pioneer neurons in the reeler cortex have not been analyzed. The tangential migration of interneurons from the ganglionic eminences via the MZ—in this case, via the superplate—is apparently not disturbed (Hevner et al. 2004). As we shall describe in more detail in Sect. 7.3.1.2, the absence of CR cells in p73-mutant mice does not produce a *reeler*-like phenotype (Fig. 18A–F), which challenges the crucial role attributed to this cell type in preplate partition and raises the possibility that other cell populations in the preplate are also important for the initial steps of corticogenesis.

7.3
Distribution of Reelin in the Developing Cortex

The reeler mouse paradigm shows that Reelin is necessary for the establishment of the inside-out migration gradient and cortical lamination. To gain insight into the first steps of human corticogenesis, we examined Reelin expression in human cortex from the earliest preplate stage at 5 GW to the end of gestation and even further into postnatal life (Meyer et al. 2000; Meyer and Goffinet 1998, Meyer et al. 2002a).

7.3.1
Cajal–Retzius Cells

The developmental history of Reelin cannot be separated from the history of CR cells, which dominate the embryonic and fetal cortex until the first Reelin-positive interneurons appear in the cortical plate by the end of gestation. The mature CR cells of the human cortex around midgestation form an impressive, highly differentiated cell population that is unmatched by their homologues in non-primate species (Fig. 19).

CR cells owe their name to Retzius, who discovered them in 1893 in the fetal human cortex, and to Cajal, who described them almost concurrently (1891–1899)

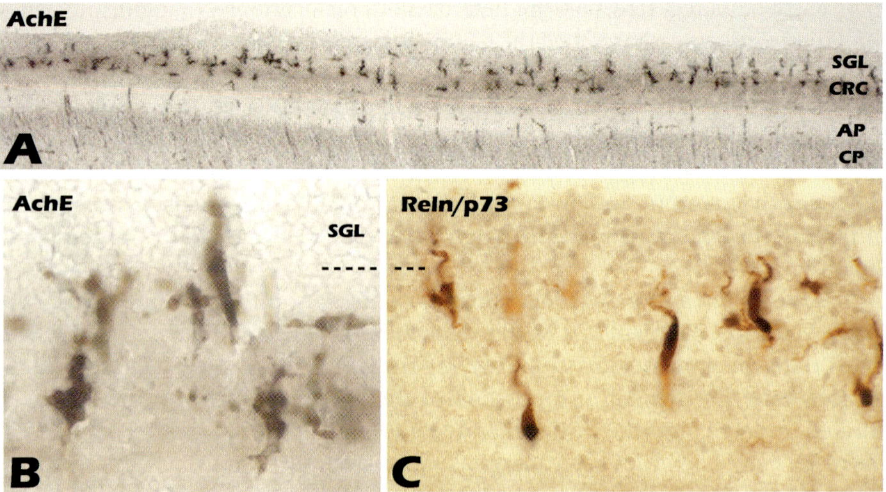

Fig. 19A–C Human CR cells at midgestation. **A** Panoramic view of CR cells in the MZ at 22 GW, stained with acetylcholinesterase (*AChE*) histochemistry. This 80-μm-thick frozen section from the temporal lobe reveals the abundance of morphological varieties, with both vertically and horizontally oriented somata. **B** Detail of vertical and horizontal AChE-positive CR cells. **C** Double immunostaining showing Reelin (*Reln*; *brown*, in the cytoplasm) and p73 (*black*, in the nucleus) in CR cells at 24 GW, in a 10-μm-thick paraffin section. At this stage, AchE, Reelin, and p72 are restricted to CR cells and not expressed in SGL cells

in rodents, lagomorphs, and human infants. Ever since these classical Golgi studies, CR cells have been the subject of many debates and speculations regarding their fate, their striking and often variable morphology, and even their neuronal or glial nature (reviewed in Meyer et al. 1999). Initially, they are closely apposed to the basal lamina, with which they may establish direct contact, but later in development, they separate from the pial surface and descend to deeper levels of the MZ (Derer 1979; Meyer et al. 1998). For almost a century, CR cells were considered as a mere curiosity, as neurons displaying a striking and unusual morphology but lacking a known function, until D'Arcangelo et al. (1995) reported the cloning of the *reelin* gene and localized the Reelin protein to the CR cells in the embryonic MZ. An independently derived antibody, CR-50, identified a cell-surface protein expressed by CR cells in the normal developing cortex but absent from reeler cortex, since the antigen is an epitope of Reelin (Ogawa et al. 1995; D'Arcangelo et al. 1997). These discoveries revived the general interest in CR cells and prompted new studies using state of the art techniques. As a result, some of the previous ideas on CR cells had to be changed. Of the many aspects of CR cells, we want to point out those ones that are relevant for human cortical development.

7.3.1.1
What Is a CR Cell?

We tried to answer this question in a previous paper (Meyer et al. 1999) where we defined CR cells as Reelin-expressing members of a heterogeneous cell family, most of which reach the MZ by tangential migration from an extrapallial source. Different subpopulations of CR cells invade the cortex at various moments of corticogenesis, and their morphological and neurochemical heterogeneity is probably responsible for the often contradictory statements in the literature. A classical marker of CR cells is AchE, which visualizes in the entire population of CR cells at midgestation (Fig. 19 A, B; Krmpotic-Nemanic et al. 1987; Meyer and Gonzalez Hernandez 1993) but is certainly not CR-cell specific. In the last years, a variety of proteins have been proposed to be characteristic of CR cells. Of these markers, Reelin is the most relevant because of its functional involvement in the regulation of neuronal migration. It is expressed at particularly high levels in CR cells (Fig. 19C, Fig. 21), but is also widely present in cortical interneurons, including small neurons in the SGL and the postnatal molecular layer, so that Reelin expression alone is not sufficient to define a CR cell. The calcium-binding protein calretinin has often been used as a CR cell marker (Fonseca et al. 1995; Del Rio et al. 1995; Schierle et al. 1997; Weisenhorn et al. 1994), but is also widely expressed in other neuronal classes in the developing cortex, for instance in the subplate and in subpial granule cells (Meyer and Goffinet 1998, Meyer et al. 2000); later in life, it is a common marker of nonpyramidal cells (e.g., Rogers 1992; Hof et al. 2000). Conversely, CR cells express also other calcium-binding proteins such as calbindin and parvalbumin (Verney and Derer 1995; Huntley and Jones 1990). The transcription factor Tbr1 is expressed in early-generated neurons of the cortex, including CR cells, subplate cells, and neurons of layer VI (Hevner et al. 2001, 2003), and thus does not allow one to identify the individual cell types of the preplate. The tumor protein p73 (Kaghad et al. 1997) is more selective since specific p73 isoforms are expressed in CR cells (Yang et al. 2000; Meyer et al. 2002a, 2004) but not in other cell types of the developing cortex (Figs. 19C, 21F). Probably the most pragmatic solution would be to identify the CR cells using a combination of neurochemical markers and morphological criteria. According to our observations, the morphologically highly characteristic CR cells of the human cortex invariably co-express Reelin and p73 (Meyer et al. 2002a).

7.3.1.2
How Necessary Are Cajal–Retzius Cells?

CR cells secrete high levels of Reelin, and are in fact the main source of this protein in the mammalian embryonic cortex. The example of the *reeler* mice clearly illustrates the consequences of the absence of Reelin secretion on cortical architecture. The Reelin signal has been proposed to undergo amplification during evolution (Tissir et al. 2002; Bar et al. 2000), and in fact the human CR cells are more numerous and more highly developed than in any other species. The structural differentiation of human CR cells extends most notably to their axonal plexus that forms a dense, compact fiber layer separating the cortical plate from the MZ (see Fig. 14, A, D; Meyer and Gonzalez Hernandez 1993). This axonal plexus is almost absent in mice, although Derer et al. (2001) showed that the axons of mouse CR cells contain secretory reservoirs of Reelin that may be important for delivering the protein into the extracellular matrix. Likewise, a secretory function of the huge human CR axonal plexus would amplify the Reelin signal and thereby increase the efficiency of the Reelin–Dab1 pathway in the human cortex. In keeping with the significance of CR cells in humans is the observation that certain human lissencephaly syndromes are associated with an extreme hypoplasia of CR cells (Fig. 24; Meyer et al. 2002b).

The phenotype of the p73-deficient mouse may shed light on the functions of CR cells, because this mutant does not develop CR cells (Fig. 18B, F; Yang et al. 2000; Meyer et al. 2002a). Whereas cortical lamination was almost not affected, the absence of hippocampal CR cells in p73-deficient mice prevented the formation of the hippocampal fissure (Fig. 20), suggesting that CR cells may be particularly important for cortical folding (Meyer et al. 2004). This aspect of CR cells is difficult to explore in the lissencephalic mouse brain, but points to an evolutionary progression of this cell population.

During evolution, CR cells may have become more important for the large, highly folded human brain than they are for the small lissencephalic mouse cortex. This would explain why the loss of CR cells in p73-knockout mice does not give rise to a *reeler*-like cortical malformation (Yang et al. 2000; Meyer et al. 2004). In p73-deficient embryonic mice, CR cells did not develop at their origin, the cortical hem, and thus did not migrate into the cortical MZ (Meyer et al. 2002a, 2004). However, the pioneer plate split into superficial and deep pioneer neurons (Fig. 18F), and MZ, cortical plate, and subplate developed almost normally, although the cortex displayed other, p73-dependent alterations such as a generalized hypoplasia. Since in the p73-deficient cortex Reelin was faintly expressed by other cell classes, we suggested that very little Reelin may be sufficient to build a mouse brain, and the high expression in CR cells might represent an excess that is not really needed. Similar findings were reported by Yoshida et al. (2006) after genetic ablation of the cortical hem. The hem-deprived mice lacked CR cells but displayed a normal cortical lamination. The experiments by Magdaleno et al. (2002) who generated *rl/rl ne-reelin* transgenic mice, which expressed ectopic Reelin in the cortical VZ under the control of the nestin promoter, showed that this ectopic Reelin was able to partially rescue some of the malformations of the *reeler* phenotype. In these animals, preplate splitting occurred normally, although lamination was severely disturbed. Altogether, these data show that Reelin in CR cells does not provide attracting or stop signals for migrating neurons, that CR cells are not needed for preplate partition, and that they are disposable for cortical lamination in mice. They also raise the question of whether the functionally relevant gene product in CR cells is Reelin, p73, or any other of the many proteins present in CR cells (Yamazaki et al. 2004). On the other hand, Reelin is widely expressed by cortical interneurons (Sect. 7.3.2), which may be responsible for many of the activities usually attributed to CR cells.

7.3.1.3
The Developmental History of CR Cells in the Human Cortex

In our studies on the expression of Reelin and p73 in embryonic and fetal human brains we showed that distinct CR cell populations emerge successively at different time points of development (Meyer and Goffinet 1998; Meyer and Wahle 1999; Meyer et al. 2000, 2002).

A few Reelin/p73-positive CR cells were present at the very beginning of corticogenesis at 5 GW, (CS 16; Fig. 21A), when there was only a neuroepithelium and a narrow marginal zone. CR cells were slightly more numerous at 6 GW (CS 17), when they were arranged in a single row in the still narrow marginal layer. At this early stage, no other neuronal classes were observed, and thus Reelin/p73-positive neurons were in fact the first neurons of the neocortex. At 6.5 GW (CS 19), Reelin-positive cells increased in number, particularly in the rostral part of the hemisphere representing the prospective frontal lobe. Here they seemed to originate from radial columns that traversed the whole width of the Reelin-negative neuroepithelium and spread into the marginal layer (Fig. 25B). Interestingly, cells in these radial columns did not coexpress p73 and may thus represent a distinct CR subpopulation that originates in the frontal pallial VZ. In general, early Reelin-positive neurons were rather small and immature and did not yet resemble the large elongate CR neurons characteristic of the foetal stages.

A dramatic change in the number and distribution of CR cells in the MZ occurred just after the appearance of the cortical plate at 8 GW (CS 21). Reelin/p73-positive CR cells suddenly increased in number, with the highest CR cell density in the

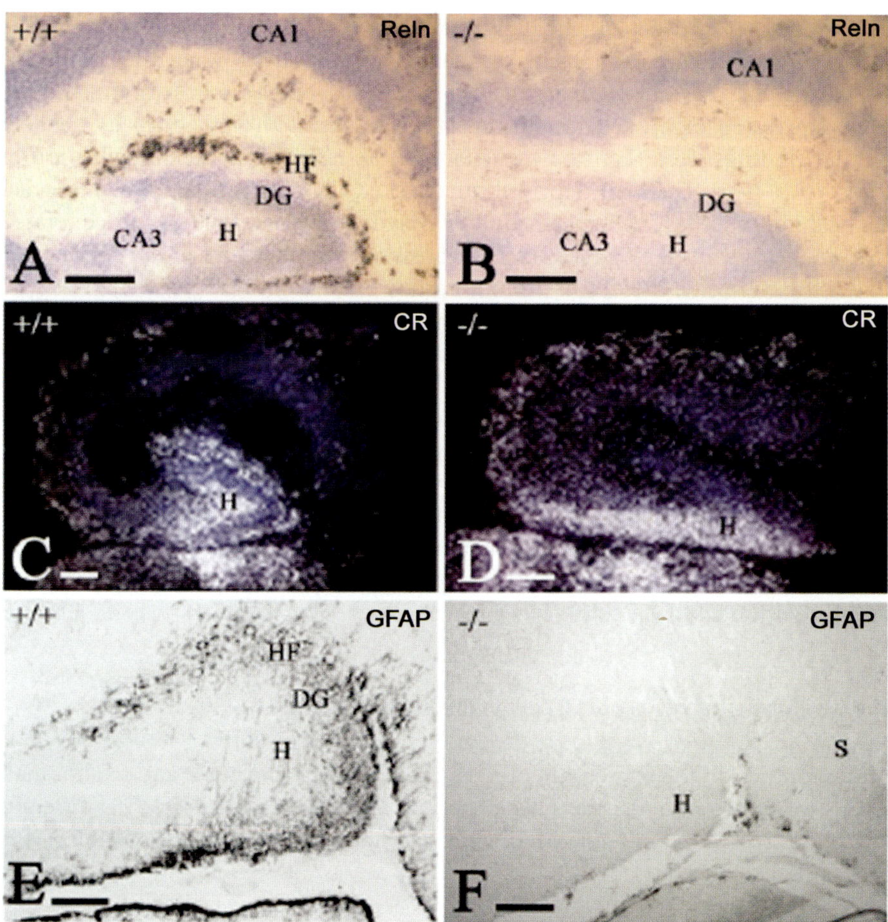

Fig. 20A–F The hippocampal fissure does not form in the absence of CR cells. **A, B** In situ hybridization (ISH) for Reelin (*Reln*) in **A** wildtype and **B** p73$^{-/-}$ mice at P2. CR cells are absent, and the molecular layers of the dentate gyrus (*DG*) and CA1/CA2 are fused. **C, D** Calretinin (*CR*) ISH in **C** wildtypeand **D** p73$^{-/-}$ mice at P2. In the mutant, calretinin-positive cells in the hilus (*H*) have abnormal distribution. **E, F** GFAP stains subpial astrocytes in the hippocampal fissure (*HF*) and the radial glia scaffold of the DG at P2. In the mutant, astrocytes are found only along the medial brain surface, since there is no hippocampal fissure. These findings suggest that p73 in CR cells may be important for cortical folding. *H,* hilus; *S,* subiculum. *Scale bars*: **A, B**, 200 µm; **C–F**, 100 µm. (Adapted from Meyer et al. 2004, with permission)

medial cortical wall. We traced the origin of this migratory wave of CR cells to the cortical hem at the interface of the choroid plexus and prospective hippocampus. The cortical hem was rather inconspicuous at earlier stages, and although it was characterized by the expression of p73 from the earliest stages of corticogenesis, only at the very precise time point of CS 21 did it give rise to large numbers of p73-positive neurons which began to co-express Reelin after their arrival in the MZ (Fig. 17 C, D). At 10 GW, after the formation of the hippocampal anlage, the cortical hem underwent regression and produced no more CR cells. The origin of CR cells in the cortical hem and their mediolateral migration into the neocortical MZ, suggested by our observations in the human, was confirmed in p73-deficient mouse embryos which failed to express p73 transcripts in the hem and lacked CR cells (Meyer et al. 2002). More recent studies in other mouse mutants were able to replicate the finding that a substantial proportion of CR cells originates in the cortical hem (Takiguchi-Hayashi et al. 2004; Muzio and Mallamaci 2005; Bielle et al. 2005; Yoshida et al. 2006).

Remarkably enough, the developmental history of human CR cells does not end with the emergence of the cortical-hem population. They increase in number even after the appearance of the cortical plate at 8 GW, keeping pace with the expanding surface of the cortex (Meyer and Goffinet 1998, Meyer and Wahle 1999); they change their shape and position in the MZ quite drastically (Fig. 21D), and finally undergo developmental cell death.

We summarize here the kinetics and developmental changes of human CR cells, observed with the above-mentioned CR-specific markers, AChE histochemistry, and DiI tracing (Meyer and Goffinet 1998; Meyer and Gonzalez Hernandez 1993; Meyer et al. 2000, 2002a). In the initial stages of cortical migration, from 5 to 13 GW, most CR cells were closely apposed to the pial surface and had a horizontal orientation and bipolar shape (Fig. 21A, B). They descended to a deeper position when the SGL appeared in the upper part of the MZ, and displayed now slender, vertically arranged somata and profuse, ascending processes (Fig. 19, 21D). Concurrently, other morphologies were observed as well, and deep horizontal somata showed strangely shaped vertical appendages. Around 23 GW, CR cells began to show signs of degeneration, such as cytoplasmic vacuoles and broken processes (Fig. 21F), and they mostly disappeared from the MZ after 30 GW, although a few residual CR cells can be observed even after birth. CR cells thus die when migration is completed and cortical lamination is established. In parallel with the disappearance of the huge CR cell population, the dense CR plexus in the deep MZ breaks down and disappears.

The mechanism of CR cell degeneration is not known, but we propose that is may be related to the expression of p73, which is a tumor protein of the p53 family (Kaghad et al. 1997). A versatile protein, p73 may appear in many isoforms, some of which have opposite functions. DeltaNp73, the dominant isoform in CR cells (Yang et al. 2000; Meyer et al. 2004), has anti-apoptotic activities (Pozniak et al. 2000, 2002; Walsh et al. 2004). However, around midgestation, CR cells also begin to express TAp73 isoforms (unpublished observations), which are known to be pro-apoptotic (Jost et al. 1997; reviewed by Yang and McKeon 2000). DeltaNp73 is expressed in the nucleus of CR cells (Fig. 21F), whereas TAp73 is also present in the cytoplasm (Fig. 21D). The degeneration and death of CR cells may be related to the complex expression pattern of both the pro- and anti-apoptotic p73 isoforms.

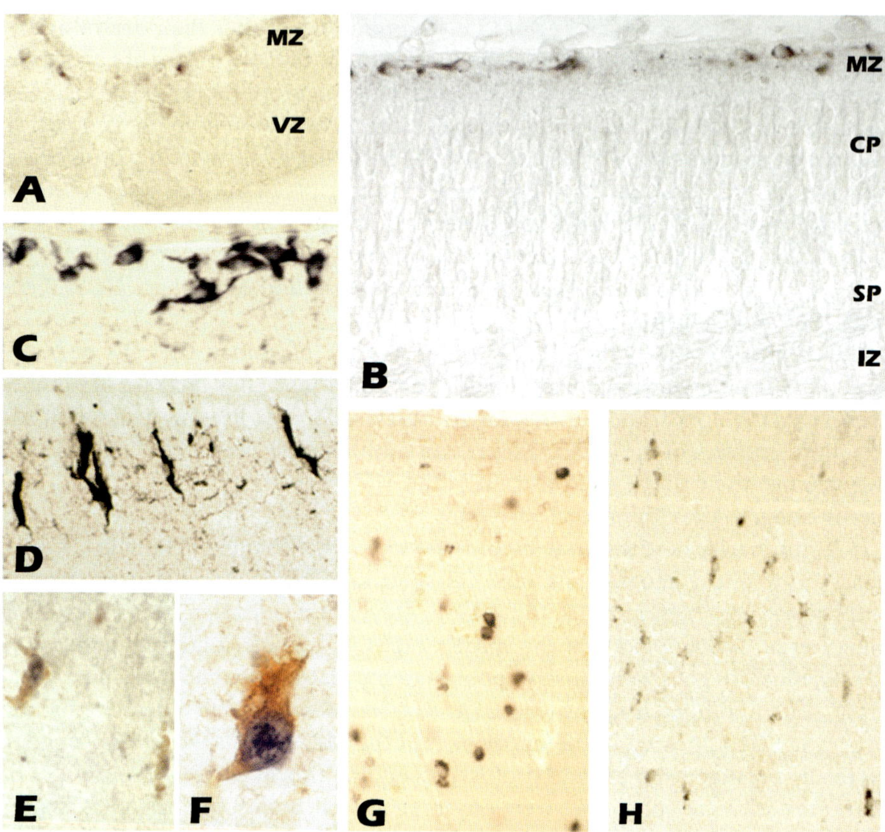

Fig. 21A–H The development of Reelin-expressing neurons in the human cortex. **A** Reelin appears very early in corticogenesis, at 5 GW. **B** At 10 GW, CR cells display a bipolar shape and subpial position. **C** At 15 GW, CR cells increase in number and structural complexity. **D** By midgestation, they assume an often vertical orientation and descend to a deeper position in the MZ. **E, F** They degenerate in the last months of gestation, showing vacuoles and broken processes, although they continue to express the specific markers p73 (*black* nucleus) and Reelin (*brown* cytoplasm). **G** After the disappearance of CR cells, the molecular layer is populated by many small Reelin-positive interneurons, which may differentiate from the SGL. **H** Reelin-positive interneurons also appear in the cortical plate at the end of gestation

The death of CR cells has been called into question repeatedly, based on the argument that they become diluted in the growing cortical surface and that occasionally a CR cell, or CR-like cell, can be found even in the adult cortex (e.g., Marin Padilla 1990; Marin-Padilla and Marin-Padilla 1982; Martin et al. 1999; Belichenko et al. 1995). Furthermore, it seems that dying CR cells cannot be detected by TUNEL staining (Spreafico et al. 1999). However, the reality of the dramatic breakdown of the entire CR cell and fiber system can be assessed by examining fetal human brains after midgestation in a systematic fashion, gestational week by gestational week, using CR-cell-specific markers and DiI tracing, as we have done in our studies.

7.3.1.4
A Timetable of the Origins of Human CR Cells

Many uncertainties and contradictions in the literature on CR cells are related to their diverse morphologies and developmental origins (reviewed by Meyer et al. 1999). According to Marin Padilla (1971, 1978, 1990), CR cells represent an ontogenetically and phylogenetically ancient cell type that remained basically unchanged during evolution, and it was assumed that observations in one species could easily be applied to others.

Numerous birthdating studies in rodents and carnivores confirmed the early generation of CR cells (e.g., König et al. 1977; Rickmann et al. 1977; Raedler and Raedler 1978; Luskin and Shatz 1985; Bayer and Altman 1990, 1991; Jackson et al. 1989). However, as pointed out above, human CR cells evolved to a much higher level of complexity than their rodent homologues, with distinct subpopulations appearing at specific time points of development. It is thus questionable whether birthdating experiments in rodents can solve the origins of human CR cells. In rodents, CR cells are born at an early preplate stage, slightly before subplate neurons (Altmann and Bayer 1990; Raedler and Raedler 1978; König et al. 1977). However, an important and usually neglected aspect is the place of birth. The concept of Marin-Padilla that CR cells are the first neurons derived from the cortical neuroepithelium has been defended by many authors, and only recently has it been accepted that most CR cells are born outside the cortical VZ and spread tangentially over the neocortex (Meyer et al. 2002; Muzio and Mallamaci 2005; Bielle et al. 2005; Yoshida et al. 2006).

Summarizing our results, we would propose the following timetable of CR cell origins:

1. A sparse early Reelin/p73-positive CR cell population is present during the initial phase of corticogenesis, and its precise origin is unknown. A subpopulation expressing Reelin but not p73 may derive from specific sectors of the pallial neuroepithelium.
2. The cortical hem population appears at the onset of the cortical plate and is characterized by the co-expression of Reelin and p73. In human, this population seems to be the predominant one.
3. After the establishment of the cortical plate and the formation of the dorsal hippocampus, CR cells still increase in number. We proposed an origin of this late population in the retrobulbar olfactory forebrain, which includes part of the septal eminence, and tangential migration from ventral to dorsal (Meyer and Wahle 1999; Meyer et al. 2002).
4. After the establishment of the SGL, smaller CR-like neurons differentiate from immature neurons in the SGL (Meyer and Goffinet 1998). Most CR-like neurons in the SGL also co-express Reelin and p73.

5. At the end of gestation, a proportion of SGL neurons differentiate into Reelin-positive interneurons of the molecular layer and perhaps even of deeper layers. They express Reelin but not p73 and have no morphological similarity with CR cells.

7.3.1.5
A Comparison of Rodent and Human CR Cells

The stereotype of a CR cell in the literature is that of a bipolar, horizontally oriented cell just beneath the pia. Rodent CR cells are in fact horizontal and bipolar, as are human CR cells during the initial stages of corticogenesis. However, from approximately 12 to 25 GW, human CR cells undergo continuous morphological changes, such as their often vertical orientation and deep position and their striking alignment at regular intervals (Fig. 19). The arrangement of their dendritic and axonal processes appears often bizarre and is unlike that of any other cortical neuron. Rodent CR cells are much less differentiated, although their developmental timetable shows certain parallels with the human model, since after birth they also descend to deeper levels of the MZ and then disappear (Meyer et al. 1998). More importantly, human CR cells give rise to a massive horizontal fiber plexus which forms the anatomical boundary between the cortical plate and the MZ. The CR plexus, which can be visualized by immunostaining for Reelin, also contains the neurofilament proteins SMI31 and SMI32 (Verney and Derer 1995). Derer et al. (2001) examined Reelin expression in axons of rodent CR cells by electron microscopy; axonal fibers were earlier found to be rather thin and smooth, without any preferential orientation (Derer and Derer 1990). The axonal boundary between MZ and cortical plate in humans may be important for the migratory behavior of cortical plate neurons, the more so as the plexus may amplify the Reelin signal in the extracellular matrix. The CR axonal plexus may be defective in cortical malformations that come with abnormalities in number and position of CR cells, such as in Miller–Dieker lissencephaly (Meyer et al. 2002), which may lead to a disruption of the cortical plate/MZ boundary and altered migration, although Reelin expression in CR cells is not affected.

Since the mouse has become the most stringent experimental instrument in current neuroscience research, it is import to recognize possible species-specific differences. CR cells exemplify a significant difference between rodents and humans. They may express the same gene products in both species; however, human developmental abnormalities may not be detectable at the level of gene expression but rather at the level of morphological differentiation (see Sect. 9.4.2).

7.3.2
Reelin in the Cortical Plate

In rodents, CR cells degenerate and die soon after birth (Del Rio et al. 1995; Meyer et al. 1998). However, Reelin does not disappear from the cortex, because it is produced by other cellular sources (Schiffmann et al. 1997; Alcántara et al. 1998). Reelin mRNA and protein are extensively expressed by GABAergic cortical interneurons, with the greatest density in layers I–III and V–VI (Pesold et al. 1998). Further characterization of the cortical interneurons was achieved by using double immunostaining and combined Golgi staining-Reelin labeling (Pesold et al. 1999). Co-expression with Reelin was often found in interneurons that express neuropeptide Y or somatostatin, but rarely calretinin or calbindin. Basket and chandelier cells, which are important categories of interneurons, express parvalbumin but not Reelin. Some Reelin-expressing interneurons display a horizontal or bitufted morphology or resemble Martinotti cells (Pesold et al. 1999). Interestingly, Reelin mRNA is expressed in the glutamatergic granule cell layer of the cerebellum, whereas Reelin protein is most abundant in the parallel fiber system and in the extracellular space, which suggests that Reelin is produced in the granule cells, transported by their axons, and finally secreted into the extracellular matrix (Pesold et al. 1998). Similarly, in the macaque cortex, Reelin was shown to be present in the neuropil and in axonal pathways and terminal arborizations, suggesting axonal transport, and a light Reelin-immunostaining was also observed in many pyramidal-shaped neurons of the monkey cortex (Martínez-Cerdeño and Clascá 2002; Martinez-Cerdeño et al. 2002). The presence of Reelin in cell somata, axonal pathways and gray matter neuropil suggested that Reelin may influence most brain circuits in the adult primate brain.

Electron microscopy studies in monkeys supported the presence of Reelin in interneurons of the adult cortex. A quantitative analysis of glutamic acid decarboxylase (GAD) 67 mRNA and Reelin mRNA showed layer-dependent co-expression profiles (Rodriguez et al. 2002). In layer I, almost every cell co-expressed GAD and Reelin, whereas the lowest rate of colocalization was in layer IV. In general, approximately 50% of GABAergic interneurons expressed Reelin or its mRNA. Differences were observed also across areas, with the visual cortex displaying more Reelin-expressing interneurons than other areas, whereas pyramidal cells did not express Reelin (Rodriguez et al. 2000, 2002). Furthermore, immunoelectron microscopy in adult monkey cortex showed that Reelin forms aggregates in the extracellular matrix in the proximity of postsynaptic domains of dendritic spines belonging to pyramidal neurons. In these postsynaptic densities, Reelin colocalized with the α3 subunit of integrin receptors, which led to the suggestion that Reelin might act as a putative endogenous ligand for integrin receptors (Rodriguez et al. 2000).

7.4
Reelin in Postnatal and Adult Human Cortex

In the postnatal and adult human cortex, immunohistochemistry using the anti-Reelin antibody 142 showed high numbers of small, intensely Reelin-positive neurons in layer I and the upper part of layer II in all neocortical areas (Fig. 21G). Relatively few, more faintly stained interneurons were also in deeper layers (Fig. 21H), and occasionally even in the white matter. Pyramidal cells were usually Reelin negative, with the exception of the entorhinal cortex. The cytoarchitecture of the entorhinal cortex differs from that of the neocortex, and one of its peculiarities is the presence of clusters of large stellate cells in layer II, which give rise to the perforant path connecting the entorhinal cortex with the hippocampus (Steward and Sconville 1976; Witter and Groenewegen 1984). In all species examined so far—rodents, carnivores, humans—the large stellate cells are moderately Reelin positive (Perez-Garcia, Drakew et al. 1998; Alcantara et al. 1998, Pesold et al. 1998). In addition, pyramidal cells of deeper entorhinal layers also express Reelin. This rather unusual expression pattern indicates that Reelin may play a role in the entorhinal–hippocampal pathways, which are important for learning and memory functions. It is also interesting to note that entorhinal layer II neurons are particularly vulnerable for degenerative alterations in Alzheimer's disease (Van Hoesen and Hyman 1990; Gomez-Isla et al. 1996). The presence of Reelin may render these neurons more sensitive to neurodegeneration, the more so as ApoE, which is a susceptibility gene for Alzheimer's disease, inhibits the binding of Reelin to VLDLR and ApoER2 (D'Arcangelo et al. 1999). The evidences for the possible relationship between Alzheimer's disease and Reelin have been summarized by Tissir and Goffinet (2003) and will not be considered here.

Another interesting aspect of Reelin in the human brain is its presence in the cerebrospinal fluid (CSF; Ignatova et al. 2004), where it seems to be derived from the brain and not from the plasma. A possible correlation between alterations of Reelin and neurological diseases has so far not been established.

In the last years, a number of studies have drawn the attention to the putative role of Reelin in certain mental diseases. A decrease of Reelin has been proposed as a putative vulnerability factor for schizophrenia (Impagnatiello et al. 1998). Measurements of Reelin mRNA content in temporal cortex, prefrontal cortex, hippocampus, and caudate nucleus in schizophrenia showed a decrease compared to control patients. The interstitial neurons of the cortical white matter also showed a decrease of Reelin mRNA in schizophrenic patients (Eastwood and Harrison 2003). Reduction of Reelin immunoreactivity in the hippocampus was described in patients with schizophrenia, bipolar disorder, and major depression (Fatemi et al. 2000). A decrease was noted also for GAD67mRNA. Similarly, Western blot analysis revealed a decrease of Reelin protein, whereas Dab1 levels were unchanged (Impagnatiello et al. 1998).

7.5
Reelin Function in Plasticity and Learning

The expression of Reelin in the adult brain, particularly in GABAergic interneurons, indicates that it may be involved in other activities unrelated to neuronal migration and laminar positioning. Although Reelin does not affect axonal growth directly (Jossin and Goffinet 2001), there is evidence that it promotes axonal branching and synaptogenesis in the entorhinal–hippocampal projection (Del Rio et al. 1997; Borrell et al. 1999). It also promotes dendritic outgrowth from hippocampal and dentate neurons in a Dab1-dependent fashion (Niu et al. 2004). Reelin accumulates on dendritic spines of human pyramidal neurons (Roberts et al. 2005) and colocalizes with integrins in postsynaptic densities of dendritic spines (Rodriguez et al. 2000), where high local concentrations of extracellular Reelin selectively outline the neuropil around the dendritic spines (Ramos-Moreno et al. 2006). The number of dendritic spines seems to be reduced in *reeler* heterozygous mice (Pappas et al. 2001; Liu et al. 2001). Dendritic spines are linked to long term potentiation (LTP), a use-dependent increase in synaptic efficiency involved in long-term memory formation in mammals (reviewed by Yuste and Bonhoeffer 2001). Mice that lack both *ApoER2* and *VLDLR* have defects in LTP induction and fear-conditioned associative learning (Weeber et al. 2002), suggesting an important role of the Reelin–Dab1 pathway in modulating synaptic plasticity and memory formation. Recent studies (Chen et al. 2005; Beffert et al. 2005) showed that Reelin modulates NMDA receptor activity in cortical neurons, and that the ApoER2 receptor is critical to this process, because it is present in the postsynaptic densities of excitatory synapses where it forms a functional complex with NMDA receptors. Reelin signaling through ApoER2 enhances LTP through a mechanism that requires the presence of amino acids in the intracellular domain of ApoER2 encoded by an alternatively spliced exon. This exon is necessary for Reelin-induced tyrosine phosphorylation of NMDA receptor subunits, and mice deficient for this exon perform poorly in learning and memory tasks (Beffert et al. 2005; D'Arcangelo 2005). Altogether, these data point to important functions of the Reelin–Dab1 pathway in adult life also, apparently unrelated to its role in migration and cell positioning.

7.6
Evolutive Aspects of Reelin

The Reelin protein sequence is remarkably conserved among vertebrates (Lambert de Rouvroit and Goffinet 1998a). The expression of Reelin has been assessed in a large variety of species such as lamprey (Perez-Costas et al. 2002), fish (Perez Garcia et al. 2001; Costagli et al. 2002), amphibians (Perez-Garcia et al. 2001), turtles, lizards, and crocodiles (Bernier et al. 1999; Goffinet et al. 1999; Tissir et al. 2003), birds (Bernier et al. 2000), rodents (Schiffmann et al. 1987; Alcantara et al. 1998), and primates including man (Martinez-Cerdeño et al. 2002, 2003; Meyer and Goffinet 1998; Abraham and Meyer 2003; Abraham et al. 2004).

The localization of the Reelin signal in the adult telencephalon is quite diverse across non-mammals, and often reflects very striking differences of pallial organization. In the everted telencephalic hemispheres of the zebrafish, Reelin is expressed in many neurons distributed in bands extending from the ventricle to the meningeal surface. In adult amphibians (*hyla meridionalis*, the Mediterranean frog), there is no cortical plate, and neurons form instead a Reelin-negative periventricular gray matter. However, a few Reelin-positive cells occupy a position just external to the periventricular gray, which is reminiscent of the position of developmental CR cells (Perez-Garcia et al. 2001). Reelin expression in reptiles is also quite different from the mammalian pattern. In the medial and dorsal cortical fields of lizard embryos, Reelin is expressed in two layers, the MZ and the subcortex, whereas the main cell layer, the representative of the cortical plate, is Reelin negative (Goffinet et al. 1999). In the adult lizard cortex, Reelin is sparse in the external plexiform layer but abundant in the internal plexiform layer, with the exception of the lateral cortex, where the main cell layer contains numerous Reelin-positive neurons (Perez-Garcia et al. 2001). In the developing turtle, there are numerous Reelin-positive neurons at the level of the medial and dorsal fields (Bernier et al. 1999). In turn, in the adult turtle (*Clemys caspica*), where the cells of the cortical plate are more loosely packed than in lizards and lie closer to the ventricle, only few Reelin-positive cells populate the external plexiform layer (Perez-Garcia et al. 2001). The example of the embryonic crocodile cortex (Tissir et al. 2003) shows that regardless of the diverse cortical architecture of adult vertebrates, a common and characteristic feature of the embryonic amniote cortex is the presence of subpial neurons expressing the characteristic CR-cell-marker proteins Reelin and p73. The comparison of Reelin expression in embryonic representatives of the main amniotic lineages suggests that the components of the Reelin signaling pathway played an important role in the acquisition of a radially oriented, laminated cortex (Bar et al. 2000; Tissir et al. 2002; see also for review Molnar et al. 2006). By contrast, the appearance of Reelin-expressing interneurons within the cortical plate seems to be a specific trait of the mammalian lineage, perhaps related to the possible roles of the protein in neuronal plasticity and memory referred to in Sect. 7.5.

7.7
Reelin Receptors

Mutant mice lacking both VLDLR and ApoER2 display a phenotype indistinguishable from that of *reeler* and *dab1-/-* mice; Reelin binds directly to the ectodomains of both receptors, and Dab1 docks to their cytoplasmic tails (Trommsdorf et al. 1999; D'Arcangelo et al. 1999; Hiesberger et al. 1999), indicating that they function as Reelin receptors and form part of the Reelin–Dab1 signaling pathway.

VLDLR and ApoER2 belong to the low-density lipoprotein (LDL) receptor family which is an evolutionary ancient family of closely related cell-surface receptors. They are characterized by five distinct domains: (1) a ligand-binding region containing cysteine-rich repeats, (2) epidermal growth-factor-type cysteine-rich repeats, (3) a β-propeller segment with YWTD domains, (4) a single membrane-spanning segment, and (5) a cytoplasmic tail containing one or more NPxY sequences (x being any amino acid; reviewed by Herz and Bock 2002). The cytoplasmic NPxY sequence is the docking site for the PI (protein interaction)/PTB (phosphotyrosine binding domain) of Dab1. VLDLR is highly expressed in heart and skeletal muscle and in the endothelial cells of major blood vessels.

VLDLR and ApoER2 function as obligate components of a pathway that transduces the Reelin signal to the adapter protein Dab1. Both proteins are expressed in or near the layers that express Reelin during the cortical migration period, and both are co-expressed with Dab1 in neurons of the cortical plate subjacent to the Cajal–Retzius cells (Trommsdorf et al. 1999; Perez-Garcia et al. 2004). In the mouse embryonic VZ, Dab1mRNA, VLDLmRNA, and ApoER2mRNA colocalize in the cortical VZ, although receptor expression is more widespread than that of Dab1, for instance in the VZ of the ganglionic eminence where Dab1 is absent (Jossin et al. 2003). Comparison between single and double *VLDLR*- and *ApoER2*-knockout mice showed that even though the individual genotypes presented subtle malformations of the hippocampus and the cerebellum, they were less severe than the combined knockout, which had a *reeler*-like phenotype. In the neocortex, *vldlr*$^{-/-}$ mutant mice had a normal marginal zone, but no clear separation of the remaining layers which showed a radial pattern. This defect was more pronounced in *apoER2*$^{-/-}$ mice where neurons were packed into tight horizontal layers and lamination was partially inverted. Interestingly, VLDLR deficiency produced the more severe defects in the cerebellum, while ApoER2 deficiency affected mostly the neocortex (Trommsdorff et al. 1999). These studies show that both VLDRL and ApoER2 are required for normal brain development, but only the absence of both receptors disrupts the Reelin–Dab1 signaling pathway.

We examined the expression of both receptors in fetal human brains using anti-VLDLR and anti-ApoER2 antibodies (Perez-Garcia et al. 2004). Figure 22 shows the main staining pattern obtained for both lipoprotein receptors. Around midgestation, highest expression was in the upper cortical plate and in large pyramidal cells of layers III and V. Interestingly, a subset of CR cells expressed VLDLR and ApoER2 together with Dab1 and Reelin (Meyer et al. 2003). The presence of all members of the Reelin signaling pathway in some CR cells suggests

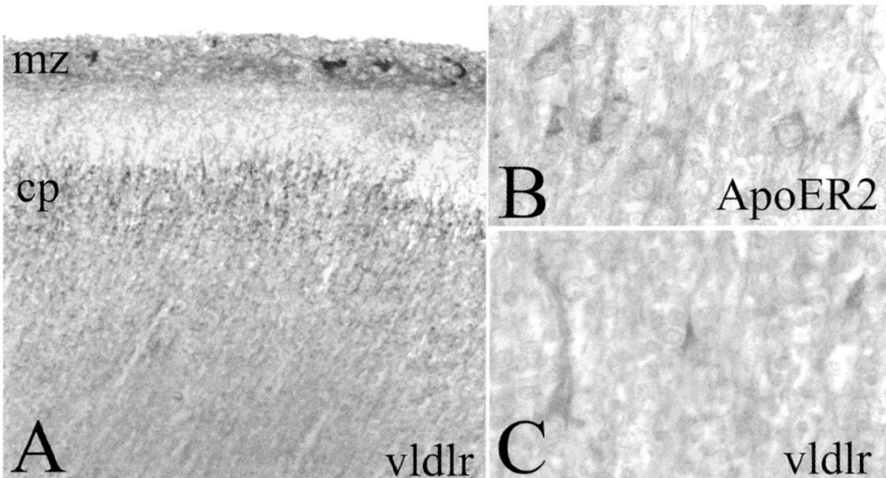

Fig. 22A–C Expression of Reelin receptors in the human cortical plate at 22 GW. Both lipoprotein receptors, **A, C** VLDLR and **B** ApoER2, are expressed in the cortical plate, most intensely in the upper tiers. VLDLP is also expressed in CR cells in the marginal zone (*mz* in **A**)

an autocrine and/or paracrine activity, perhaps related to the positioning of CR cells at regular intervals. The prominence of the Reelin–Dab1 pathway in human CR cells may be related to the evolutionary increase in number and differentiation of this cell type in the human brain.

More recently, Boycott et al. (2005) identified a deletion of the entire *VLDLR* gene as being responsible for another autosomal recessive form of cerebellar hypoplasia with cerebral gyral simplification in a Hutterite population. The affected individuals showed a simplification of the gyral pattern and a thickening of the cerebral cortex, accompanied by moderate to severe mental retardation, epilepsy and ataxia, the latter possibly related to the severe hypoplasia of the cerebellar hemispheres. On the whole, *VLDLR* deficiency has milder consequences in the cortex than *Reelin* deficiency in both mice and humans, although the cerebellar defect appears more marked.

7.7.1
Integrins—Putative Co-receptors?

An important question is whether the lipoprotein receptors are the only surface proteins that interact with Reelin. It has been proposed that members of the integrin family form part of the Reelin signaling pathway. These ubiquitous cell-surface proteins are a major mediator of cell–cell and cell–extracellular matrix interactions. There are at least 18 α and 8 β subunits that can form more than 20 different integrin receptors, which are expressed in complex and highly dynamic patterns in the developing cortex (reviewed by Schmid and Anton 2003). Reelin has been shown to interact with $\alpha3\beta1$ integrin, which suggested that it could inhibit neuronal migration by stimulating detachment from the radial glia substrate (Dulabon et al. 2000). Single mutations of the different subunits, however, did not show a *reeler*-like phenotype. Mutations in the $\alpha3$ subunit induced neuronal heterotopias in the intermediate zone, premature transformation of radial glia into astrocytes and impaired neuron–glia interaction (Anton et al. 1999). On the other hand, brain-specific inactivation of $\beta1$ integrins caused a phenotype reminiscent of cobblestone or type-2 lissencephaly, suggesting that these integrins regulate the anchorage of radial glial endfeet to the basal lamina (Graus-Porta et al. 2001). Interestingly, in $\beta1$-deficient mice the CR cells were arranged irregularly with alternating gaps and heterotopic clusters suggesting that $\beta1$ regulates the interactions between CR cells, glia endfeet and the basement membrane. Furthermore, αv-containing integrins can also bind Reelin and thus may partially be able to compensate for the $\alpha3$ defect (Anton et al. 1999).

The recent demonstration that $\alpha3\beta1$ integrin binds to the N-terminus of Reelin, while the cytoplasmic domain of $\beta1$ integrin containing the NPxY motif can bind directly with the PTB domain of Dab1, has confirmed the interaction between Reelin signaling and $\alpha3\beta1$ integrin (Schmid et al. 2005). Interestingly enough, the N-terminal domain of Reelin that associates with $\alpha3\beta1$ integrin does not overlap with the region of Reelin known to associate with VLDLR and ApoER2 (Benhayon et al. 2003; Jossin et al. 2003), raising the question of whether the binding of distinct regions of Reelin with different receptors may subserve different functions during cortical development. Reelin could potentially induce $\alpha3\beta1$ integrin receptor clustering, as has been shown for ApoER2 and VLDLR (Strasser et al. 2004), and thus modulate changes in the adhesive properties required by the migrating neurons at the end of their route.

7.8
Dab1

7.8.1
Dab1 Gene and Protein

The *Disabled-1* (*Dab1*) gene is another key component of the Reelin signaling pathway in mouse and human (Howell et al. 2000; Rice et al. 1998; Rice and Curran 2001). The Dab1 gene maps to mouse chromosome 4, and human chromosome 1p32-p31 (Lambert de Rouvroit and Goffinet 1988b). Inactivation of Dab1 by homologous recombination (Howell et al. 1997) or by spontaneous mutations in scrambler or yotari mutant mice (Sheldon et al. 1997; Ware et al. 1997) produce a phenotype similar to that of Reelin-deficient mice. There is no additional defect in mice lacking both Reelin and Dab1, which indicates that the two proteins form part of a linear signaling pathway (Howell et al. 1999). At present, no human disease caused by mutations of the *Dab1* gene has been identified.

The genomic organization of the *Dab1* gene is highly complex in both human and mouse. The gene extends over more than 1Mbp of genomic DNA owing to the presence of large introns and several alternative transcription initiation sites. The presence of alternative promoters and alternative transcription initiation sites indicates a tight regulation (Bar et al. 2003).

The main Dab1 protein is 555 amino acids long. Alternative promoter usage, alternative splicing and polyadenylation give rise to several Dab1 isoforms, the expression patterns and functional relevance of which has not yet been determined. The 180-amino acid N-terminal PI/PTB domain docks to the short cytoplasmic tail of VLDLR or ApoER2 at the level of the NPXY motif, with a preference for unphosphorylated NPxY sequences (Howell et al. 1999b). The binding of Reelin to the extracellular part of both lipoprotein receptors induces phosphorylation of tyrosine residues of Dab1, in particular Tyr^{198} and Tyr^{220} (Keshvara et al. 2001), and the main kinases involved are Fyn and Src (Bock and Herz 2003; Arnaud et al. 2003). After tyrosine phosphorylation, Dab1 is ubiquitinated and degraded by the proteasome (Arnauld et al. 2003). The PTB domain and tyrosine phosphorylation are essential for Dab1 functions, because mice expressing mutated forms of Dab1 protein in which 5 important Tyr residues were replaced by Phe, have a *reeler*-like phenotype (Howell et al. 2000). Tyrosine phosphorylation of Dab1 is, however, not sufficient for Dab1 function, and the C-terminal region contains consensus serine/threonine phosphorylation sites which are substrates of cyclin-dependent kinase 5 (cdk5)/p35 kinase (Keshvara et al. 2002). Defective Reelin signaling results in Dab1 hypophosphorylation along with a drastic upregulation of Dab1 protein levels (Hiesberger et al. 1999; Howell et al. 1999).

The construction of $Dab1^{+/+} \leftrightarrow Dab1^{-/-}$ chimeric mice addressed the question of whether Dab1 activity is cell autonomous or whether Dab1 activation produces non-cell autonomous effects, for instance by regulating the production of surface or extracellular molecules that may affect the behavior of other cells (Hammond et al. 2001). $Dab1^{+/+}$ cells were capable of migrating radially and arranging themselves in a columnar fashion in a $Dab1^{-/-}$ environment. $Dab1^{+/+}$ cells migrated to the superficial part of the mutant cortex forming a multilayered "supercortex" with correct inside-out layering. Most $Dab1^{+/+}$ neurons expressed the transcription factor Emx1, which is a marker of pyramidal cells, and gave rise to extensive projections into the underlying white matter. The cell-autonomous function of Dab1 was indicated by the fact that $Dab1^{+/+}$ cells failed to rescue the inversion of cortical layers and the abnormal structure of the MZ formed by $Dab1^{-/-}$ cells. Interestingly enough, GABAergic interneurons, known to derive from ganglionic eminences and to migrate tangentially, were underrepresented in the $Dab1^{+/+}$ supercortex, which raises the question of whether Reelin signaling acts principally on pyramidal cells whereas interneuron migration is rather independent of Reelin.

7.8.2
Dab1 in Human Cortical Development

We analyzed the expression patterns of Dab1 mRNA and protein in human cortical development from preplate stages to term (Meyer et al. 2003). The first weak signal of Dab1mRNA was detected at CS 20 (7 GW), an advanced preplate stage. This signal appeared in the first condensation of pioneer cells in the loosely organized preplate close to the striato-cortical angle. At this stage, Dab1 protein expression in the telencephalon was faint compared to that in diencephalon and brain stem. Dab1 was also highly expressed in extracerebral ganglia such as the ganglion terminale. The pioneer plate of CS 20 (7 GW) showed intense positivity. Dab1 mRNA and protein levels were also high in the cortical plate at CS 21/22 (8/9 GW). The highest Dab1 protein signal was in the apical tips of the radially oriented somata (Fig. 23A).

Dab1 expression remained high during the period of maximum migration into the cortex, between 12 and 20 GW, with the most intense signal in the upper cortical plate, just below the MZ (Fig. 23B). Most positive cells displayed a pyramidal phenotype. After midgestation, Dab1 levels decreased, first in the deep layers, and later also in the superficial layers, and were practically undetectable at birth.

Comparison of the expression patterns of Reelin and Dab1 shows that the onset of Dab1 expression is slightly delayed with respect to the Reelin signal. This is not surprising since Dab1 has been related to the radial migration of the cortical plate. In fact, the emergence of the pioneer plate at CS 20 may be the first event in radial migration, because, as explained above, the cell populations in the preplate may arrive through tangential migration. On the whole, the two proteins were expressed in a complementary manner, with Reelin being concentrated in the MZ, and Dab1

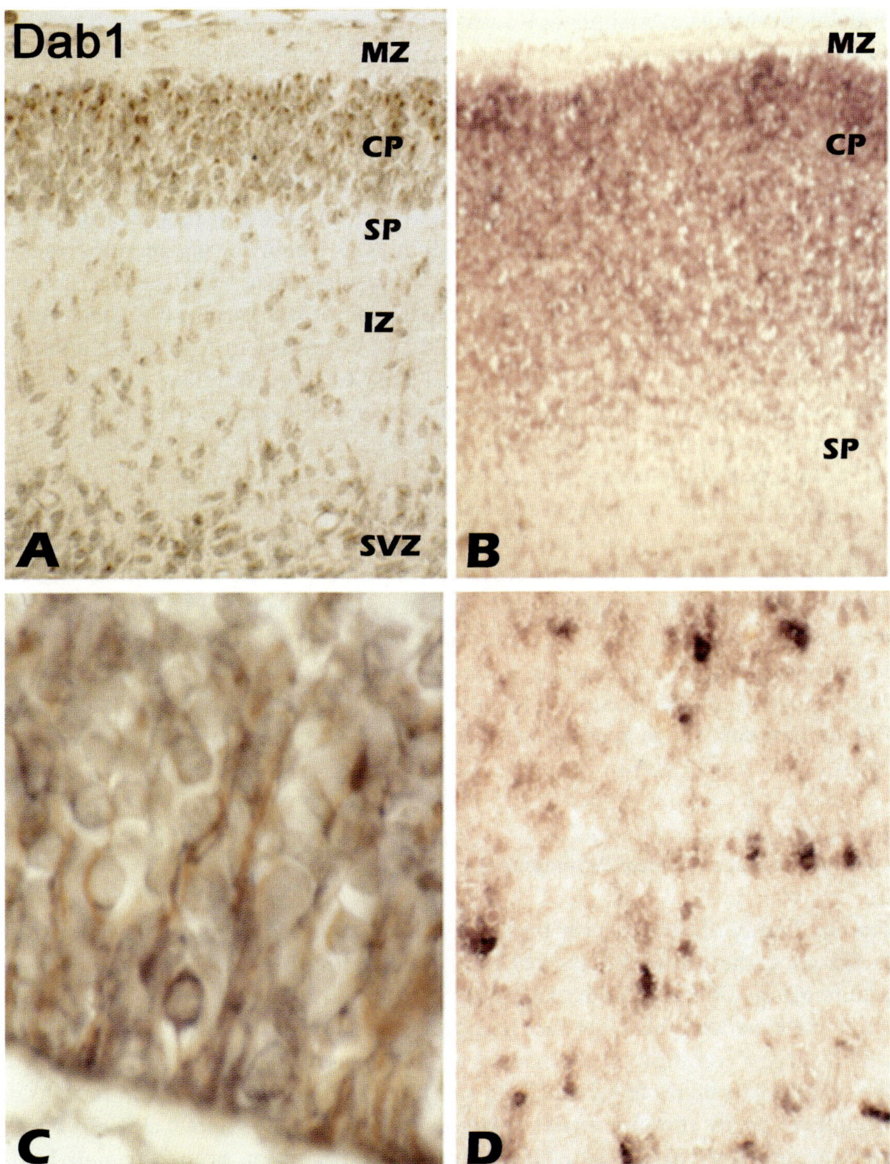

Fig. 23A–D Dab1 expression in human cortical development. **A** Dab1 protein, at 9 GW, is expressed most strongly in the cortical plate (*CP*), but also in the subplate (*SP*) and intermediate zone (*IZ*). **B** At 14 GW, Dab1 mRNA is strongly expressed in the CP, most notably in its upper tiers. **C** At 15/16 GW, Dab1 protein is in radial glia cells and fibers in the VZ. **D** At midgestation, Dab1 mRNA is also prominent in large neurons of the IZ which may represent resident cells of this compartment

in the underlying cortical plate. However, Dab1 was expressed also in places that were far removed from the Reelin signal in the MZ, namely in the subplate and the IZ. Large scattered cells in the IZ expressed high levels of Dab1 mRNA and protein, although most cells in the IZ were Dab1 negative (Fig. 23D).

Another exception to the complementarity of Reelin and Dab1 was the co-expression of Reelin and Dab1 in a subpopulation of CR cells. Since a conspicuous feature of CR cells is their regular spacing, a hypothesis would be that CR cells might control their own position through the Reelin–Dab1 pathway via an autocrine or paracrine mechanism, the more so as CR cells may also express the Reelin receptors.

A functionally significant finding is expression of Dab1 in the VZ. Dab1 transcripts and protein were observed from 12 GW onward in a subpopulation of cells in the VZ. These cells were large and close to the ventricular surface. Dab1 immunoreactivity was often concentrated at the apical (ventricular) pole of the cytoplasm, or at the endfeet of those cells that were more separated from the ventricle. At 14–16 GW, Dab1-positive somata became more numerous and localized to the SVZ, while Dab1 immunoreactivity was clearly detected in radial glial processes (Fig. 23C). The Dab1 signal was particularly high in the SVZ around midgestation, when the SVZ becomes the main proliferation compartment. Co-staining with vimentin demonstrated that Dab1 and vimentin are colocalized in a subpopulation of radial glia. Since radial glia are neuronal precursor cells (Malatesta et al. 2000; Miyata et al. 2001; Tamamaki et al. 2001; Noctor et al. 2002), the presence of Dab1 in radial glia directly implicates the Reelin–Dab1 signaling pathway in neurogenesis. Our human data are in keeping with similar observations in the rat (Luque et al. 2003).

7.9
Tyrosine Kinases in Cortical Development

The Src family of nonreceptor protein tyrosine kinases is highly conserved over metazoan evolution and plays key roles in signaling pathways that regulate cell proliferation, differentiation, and motility (Brown and Cooper 1996). In the developing mammalian brain, there are least three Src-family kinases: Src, Fyn, and Yes, which share a common mechanism of activation and can phosphorylate many of the same substrate proteins in cells in a redundant manner. Targeted mutations of *src* and *yes* do not produce an obvious brain phenotype (Soriano et al. 1991; Lovell and Soriano 1996; Kuo et al. 2005). By contrast, disruption of the *fyn* gene leads to significant brain defects, such as inappropriate laminar position and scattering of early- and late-born cortical neurons consistent with incomplete preplate splitting, although the marginal zone is clearly developed (Yuasa et al. 2004; Kuo et al. 2005). The combined deletion of *src* and *fyn* gives rise to a phenotype very similar to that of *reeler* and *scrambler* (Kuo et al. 2005). These double mutants present a hypercellular marginal zone resembling the *reeler* superplate, and lamination is disturbed. Nevertheless, the cortical malformation seems to be less severe than that of *reeler*, because there is a residual subplate and the cortical plate is not completely inverted, suggesting that other Src family members are able to maintain a minimum tyrosine phosphorylation of Dab1.

8
The Cdk5/p35 Pathway

Cyclin-dependent kinase 5 (Cdk5) will be addressed here because it belongs to a distinct molecular pathway that acts in neuronal positioning in parallel with the Reelin–Dab1 signaling pathway (reviewed by Oshima and Mikoshiba 2002), and the cortical malformation of Cdk5-deficient mice is very similar to the *reeler* phenotype. Cdk5 is a member of the cyclin-dependent serine/threonine kinase family. It has high homology to other Cdks, but is unique because it does not associate with cyclins and does not regulate the cell cycle. It is expressed ubiquitously, but displays its kinase activity mainly in postmitotic neurons of the developing and adult brain (Tsai et al. 1993). The association of Cdk5 with a neuron-specific regulatory subunit (p35 or p39) is necessary for kinase activity (Tsai et al. 1994; Tang et al. 1995). Cdk5 phosphorylates a large number of substrates including cytoskeletal components such as tau and MAP1B, other kinases such as the Rak effector Pak1 (Nikolic et al. 1998), and many other proteins including Dab1 (Keshvara et al. 2002), NUDEL (Niethammer et al. 2000), and DCX (Tanaka et al. 2004). Cdk5/p35 is thus involved in many activities, for instance in the control of the actin skeleton and microtubule dynamics, in nucleokinesis, and in neuronal migration (Xie et al. 2003). $Cdk5^{-/-}$ mice have a defect of radial cortical migration that shares some common features with the *reeler* phenotype, but is distinct because migration is interrupted at an early stage of the cortical plate (Oshima et al. 1996; Gilmore et al. 1998). The formation of the preplate is not disturbed, and early-born neurons migrate successfully and form an initial cortical plate. However, later-born neurons fail to migrate beyond their predecessors and settle beneath the subplate in an inverted fashion, from outside to inside, giving rise to an abnormal layer named the "underplate". As a result, the cortex shows a four-tiered structure consisting of a normal marginal zone, a small cortical plate, a subplate, and an underplate. This phenotype has been compared to the human 1 type lissencephaly cortex, where conventionally four layers are likewise distinguished (see Sect. 9.2). In fact, doublecortin (DCX), one of the causative genes, mutations of which give rise to type 1 lissencephaly, is a substrate of Cdk5 (Tanaka et al. 2004; Graham et al. 2004), and serine phosphorylation of DCX by Cdk5 may influence the migratory behavior of neurons through modulation of the microtubule cytoskeleton.

Inactivation of *p35* causes a similar although less severe phenotype (Chae et al. 1998; Kwon and Tsai 1998). The Cdk5/p35 complex is thus important for cortical migration, which has raised the question of whether Cdk5/p35 and Reelin–Dab1 are part of a common signaling cascade. Combined mutations of both pathways produced more extensive malformations than single *Reelin/Dab1* or *Cdk5/p35* mutations, indicating that both pathways may contribute synergistically to neuronal positioning in the developing brain (Ohshima et al. 2001). Cdk5 phosphorylates Dab1 in vitro and in vivo on serine 491, independently of Reelin, during both embryonic and adult stages, whereas tyrosine phosphorylation of Dab1 through Reelin signaling is restricted to embryonic and early postnatal ages (Keshvara et al. 2002). These results show that Cdk5/p35 and Reelin–Dab1 belong to two distinct signaling pathways that control neuronal migration, although Dab1 may serve as a point of convergence and be involved in more than one pathway.

Interestingly, Cdk5/p35 and Reelin–Dab1 also cooperate in adult life in modulating synaptic transmission, possibly by functioning synergistically and converging at a point downstream of Dab1 and Akt, a substrate of Cdk5 (Beffert et al. 2004). Cdk5 acts directly on NMDA receptors and is required for LTP induction (Li et al. 2001); likewise, $p35^{-/-}$ mice have deficiencies in spatial learning and memory, demonstrating that p35-dependent Cdk5 activity is important for learning and synaptic plasticity (Oshima et al. 2005).

9
LIS1 and DCX: Key Genes for Neuronal Migration and Cortical Folding

9.1
Human Lissencephaly Syndromes

The elaborate sulcation/gyration pattern of the human cortex is often explained in terms of an increased cortical surface that serves as the substrate for the unique cognitive abilities of our species. The large, highly convoluted human cortex can be conceived as the product of multiple, independent, and superimposed modifications of structures and developmental pathways that occurred during evolution (Caroll 2003). It may appear paradoxical that mutations of certain key genes leading to single amino acid substitutions in their protein sequence are able to completely abolish cortical folding and profoundly disturb brain architecture, wiping out in one single step the evolutionary progress. Among the genes required for the development of the convoluted, six-layered neocortex, the most salient examples are *LIS1* and *DOUBLECORTIN* (*DCX* or *X-LIS*), because their deficiency is the cause of most cases of classical or type 1 lissencephaly and the related double-cortex syndrome.

The term "lissencephaly" (smooth brain) encompasses a spectrum of malformations characterized by a simplification or complete loss of the gyri and sulci. Agyria represents the most severe form, the almost complete loss of convolutions, whereas pachygyria refers to gyri that are reduced in number but increased in thickness (Aicardi 1991; Harding et al. 1996). Cortex and hippocampus are typically affected, and often the patients also present inferior olivary heterotopia, cerebellar cortical dysplasias, and corticospinal abnormalities. Subcortical band heterotopia (SBH) or "double cortex" is also included in the lissencephaly spectrum. The advances in brain imaging, particularly MRI, and molecular genetics have allowed identification of the causative gene defect and the corresponding brain phenotype of several types of lissencephaly (Guerrini and Filippi 2004; Clark 2004; Forman et al. 2005; Chang and Walsh 2003). Classical or type 1 lissencephaly falls into the category of a neuronal migration disorder, whereas type 2 lissencephaly is rather a disorder of the basement membrane, as discussed elsewhere (Francis et al. 2006).

9.2
Type 1 Lissencephaly

In type 1 lissencephaly, the defect of cortical folding is linked to a migration defect, which leads to abnormal positioning of postmitotic neurons and a profound lamination disorder of the cortical plate. As a result, the cortex is abnormally thickened and can roughly be subdivided into four layers, which are, however, not comparable to the six layers of the normal neocortex. Only layer 1 corresponds to the marginal zone/molecular layer of the normal cortex. In a foetus with Miller–Dieker syndrome (see below)), the MZ contained decreased numbers of Cajal–Retzius cells, although Reelin expression was not affected (Fig. 24A, B). Similarly, a dramatic decrease of DCX-positive cells in the SVZ (Fig. 24C, D) indicated that this cell compartment was severely altered (Meyer et al. (2002b).

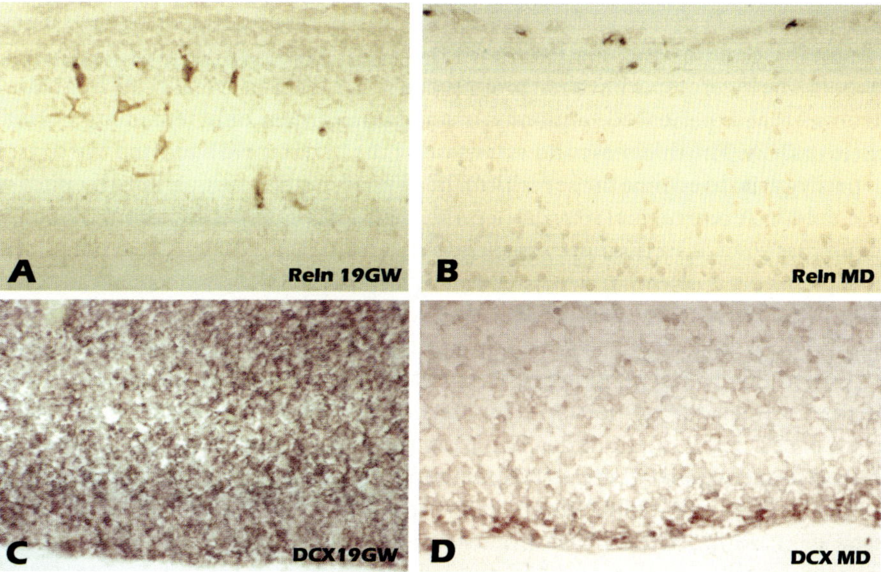

Fig. 24A–D Reelin and DCX in Miller–Dieker (*MD*) syndrome at 19 GW. **A** Reelin-positive CR cells in a normal control case, displaying the variable morphologies characteristic of this age group. **B** CR cells in MD syndrome are severely altered in number and shape, although Reelin expression is preserved. **C** DCX in VZ/SVZ in the control fetus, when DCX-positive neurons are particularly prominent. **D** DCX in MD syndrome in VZ/SVZ, showing a dramatic decrease in this cell population, although DCX expression is not altered. (Adapted from Meyer et al. 2002, with permission)

A second layer, layer 2, contains large, often disorganized pyramidal cells. Layer 3, beneath the pyramidal layer, is a narrow cell-poor zone containing myelinated fibers. The deep layer 4 is a thick, cell-dense band populated by medium-sized and small neurons (Dobyns et al. 1996; Jellinger and Rett 1976). The profound alteration of cortical architecture is incompatible with normal intellectual ability, and most patients suffer severe mental retardation, epilepsy, marked hypotonia, and lack visual contact and speech. Most children have epilepsy, which in about 75% begins before 6 months. Although some patients with severe lissencephaly may live more than 20 years, most do not (Palmini et al. 1991).

The most common cause of classical lissencephaly is a disruption of the *LIS1* gene on chromosome 17p13 (Ledbetter et al. 1992; Reiner et al. 1993). Most affected patients have spontaneous heterozygous deletions of *LIS1*, or larger deletions that encompass adjacent genes causing a syndrome with characteristic craniofacial abnormalities known as the Miller–Dieker syndrome (Dobyns et al. 1993; Reiner et al. 1993). Genetic studies of patients with type 1 lissencephaly revealed mutations in a single LIS1 allele (LoNigro et al. 1997; Chong et al. 1997) and reduction of levels of LIS1 protein (Mizuguchi et al. 1995; Fogli et al. 1999).

Mutations of the *DCX* gene on Xq22.3-q23 give rise to type 1 lissencephaly in males and subcortical band heterotopia (SBH) or "double cortex" in females (Des Portes et al. 1998a, b; Gleeson et al. 1998). In SBH, the gyral pattern is normal or simplified (pachygyria), and the defining feature is the presence of a bilateral and symmetric band of gray matter within the white matter underlying the cortical ribbon. The heterotopia has been proposed to result from an arrest of the migratory neurons that experienced random X inactivation of their only remaining normal allele of *DCX*. The thickness and extension of the heterotopic band and the degree of pachygyria determine the severity of the clinical manifestations, mostly epilepsy and mild to severe mental retardation, although some females have normal intelligence (Barkovitch and Kuziecky 2000; Dobyns and Truwit 1995). Most patients have epilepsy and it is often intractable (Bernasconi et al. 2001). Lissencephaly and SBH have been found to coexist in the same brain, and are thus considered to be part of a large agyria-pachygyria-SBH spectrum of malformations (Pinard et al. 1994).

Lissencephaly and SBH caused by mutations of *LIS1* and *DCX* have distinctive anatomical features that are useful for diagnosis and the orientation of genetic testing. Cortical malformations due to *LIS1* mutations are more severe in the parietal and occipital regions, whereas *DCX* mutations produce the reversed gradient, with malformations being more severe in the frontal cortex (Pilz et al. 1998; Gleeson et al. 2000). *REELIN*-mutations give rise to a milder pachygyria (see Sect. 7.2), and thus do not fall in the category of type1 lissencephaly.

9.3
LIS1

LIS1, also known as PAFAH1B, is widely expressed in all tissues. In the brain, it is present in dividing as well as in migrating cells, and may thus be involved in many developmental events. The fundamental role of *LIS1* in brain development is reflected by the fact that homozygous mutations are lethal in humans and in mice. Large deletions of *LIS1*, detectable by cytogenetics, are the most common mutations observed in patients with lissencephaly (Pilz et al. 1998). *LIS1* encodes a ubiquitously expressed 45 kDa protein with seven WD40 repeats (Reiner et al. 1993). The WD40 repeat is a structural motif present in a variety of proteins and thought to mediate protein–protein interactions.

Several apparently distinct roles for LIS1 in neuronal migration have been proposed. In the first place, LIS1, or PAFAH1B, is the non-catalytic subunit of the brain isoform of platelet-activating factor acetylhydrolase (PAFAH) isoform 1b, an enzyme that catalyses the inactivation of platelet activating factor (PAF; Hattori et al. 1994). However, the significance of PAF in neuronal migration is as yet unknown, and there is no apparent relationship between this function and lissencephaly.

Second, LIS1 is similar to the nuclear distribution (Nud) protein F (Xiang et al. 1995) and interacts physically and biochemically with the mammalian orthologue of NudC and with mNudE and mNUDEL during neuronal migration in vivo (Aumais et al. 2001). In the slime mold *Aspergillus nidulans*, Nud proteins are required for nuclear translocation after nuclear division, when the daughter nuclei move into the developing germ tube and the cells become multi-nucleated. The uniform distribution of the nuclei is important for correct growth and development, and mutations of the *Nud* genes lead to a failure of nuclear migration. The similarities between the NudF nuclear migration protein of *A.nidulans* and LIS1 suggested that nuclear migration is also a feature of brain development (Morris 2000). In mammals, interactions between both mNudE and NUDEL and LIS1 seem to be important for brain development. Missense mutations in *LIS1* that disrupt human cortical development produce stable proteins that fail to bind mNudE (Feng et al. 2000). This suggests that LIS1 may have a function similar to NudF and that nuclear migration (nucleokinesis) plays an important role in neuronal migration (Morris et al. 1998).

Third, LIS1 regulates the dynein motor system in mammals. Cytoplasmic dynein and dynactin are conserved throughout eukaryotic evolution. They are involved in various forms of intracellular motility, such as the separation of the spindle poles during mitosis, retrograde axonal transport, protein sorting, and organelle redistribution. Overexpression of LIS1 has profound effects on mitotic progression, mitotic spindle orientation, and chromosome attachment, and on the subcellular localization of cytoplasmic dynein and dynactin, suggesting that a LIS1–dynein interaction may be important for cell division and cytokinesis during the early stages of brain development (Faulkner et al. 2000).

Analysis of different *Lis1*-knockout mice provided valuable information about the roles of the protein in brain development (Hirotsune et al. 1998; Cahana et al. 2001; Gambello et al. 2003). As in human, *Lis1* homozygotes were not viable, and the heterozygotes showed dose-dependent severity of the phenotype, without reaching the dramatic phenotype of the human heterozygotes. The development of radial glia, CR cells, and preplate appeared unaffected, whereas dose-dependent defects were noted in the subplate. In the VZ, interkinetic nuclear migration and mitotic division were abnormal, and cell death was increased, leading to progressive thinning of the cortex and ventricular zone (Gambello et al. 2003). Abnormal tangential migration of interneurons was reported in *Lis* (+/–) embryos (McManus et al. 2004). A more recent study using in utero electroporation of LIS1 small interference RNA and short hairpin-dominant negative LIS1 in rats produced dramatic developmental defects (Tsai et al. 2005). Multipolar postmitotic cells failed to convert into the bipolar migratory stage (see Sect. 4.1) and accumulated in the SVZ, unable to resume glia-directed radial migration. In the VZ, interkinetic nuclear movement was abolished and the number of mitotic figures reduced.

9.4
DCX

Doublecortin (DCX) is responsible for the double-cortex malformation or SBH in heterozygous carrier women, and for X-linked lissencephaly in hemizygous males (Gleeson et al. 1998; des Portes et al. 1998). Most cases of SBH are caused by mutations of the DCX gene (Gleeson et al. 2000), and a minority by mutations of the LIS1 gene and other genes (Sicca et al. 2003). The DCX gene encodes a 361 amino acid protein that associates with and stabilizes microtubules (Francis et al. 1999; Gleeson et al. 1999; Horesh et al. 1999). DCX consists of an evolutionary conserved, tandemly repeated domain in the N-terminal part of the protein (the DC repeat), and a C-terminal serine/proline-rich domain (Sapir et al. 2000; Taylor et al. 2000). Each repeat alone is able to bind tubulin, but both are required to mediate microtubule stabilization, and the majority of human missense mutations cluster in the DC repeats (Taylor et al. 2000). Mutations of *DCX* may interfere with DCX-mediated microtubule polymerization. Serine phosphorylation of DCX by Cdk5 may regulate neuronal migration through an effect on microtubule dynamics (Tanaka et al. 2004; Graham et al. 2004), which indicates that DCX function is regulated by several molecular pathways. The mouse DCX is 99% identical to the human counterpart, and in both species DCX is highly expressed in the adult frontal lobe (Sossey-Alaoui et al. 1998).

The fact that mutations of both human *LIS1* and *DCX* genes produce a similar phenotype raises the question whether they interact on a molecular level. LIS1 interacts with polymerized microtubules, binds tubulin, and inhibits microtubule catastrophe events (Sapir et al. 1997, 1999); LIS1 and DCX enhance tubulin polymerization in an additive manner, and cross-talk between them may be important for microtubule function in the developing cortex (Caspi et al. 2000). A complex model of a possible interaction between LIS1 and DCX in neuronal migration, involving coupling between the centrosome and the nucleus through DCX, LIS1, and the dynein/dynactin complex, has been proposed by Tanaka et al. (2004).

In addition to its role in microtubule dynamics, DCX has been reported to have other activities: it may be involved in vesicle trafficking (Friocourt et al. 2001) and microtubule–actin crosstalk (Tsukada et al. 2003, 2005), and it seems to be enriched at the extremities of neuronal processes (Francis et al. 1999; Friocourt et al. 2005; Schaar et al. 2004).

An additional protein containing DC repeats is named DCLK (doublecortin-like kinase), also known as KIAA0369 and DCAMKLI (Burgess et al. 1999; Lin et al. 2000). The gene maps to human chromosome 13q12.3, where no neurological mutations have been described thus far. The DCLK protein contains, in addition to the DC domain, a transmembrane domain and a calmodulin kinase domain and has varying levels of expression throughout development and adulthood through the regulated expression of multiple splice variants. DCLK and DCX show similar microtubule-stabilizing activities (Burgess and Reiner 2000; Lin et al. 2000) and may play partially redundant roles in brain development. DCLK may be able to compensate the loss of DCX in *DCX*-deficient mice, although not in human *DCX* mutations. Recent rodent studies pTrovide evidence for a complex interplay of DCLK, DCX, Lis1, and dynein (Shu et al. 2006; Deuel et al. 2006; Koizumi et al. 2006). Unlike DCX, DCLK is expressed in dividing progenitor cells and plays a role in mitotic spindle formation and the regulation of cell fate determination. Furthermore, DCLK polymerizes microtubules in a LIS1- and dynein-dependent fashion, and its N-terminus plays a role in mitotic arrest at prometaphase (Shu et al. 2006). Defects of DCLK have thus far not been demonstrated in human brain.

9.4.1
DCX in Mice

DCX expression in mice is observed from E 11.5 in postmitotic neurons, but not in proliferating cells in the VZ. It is also expressed in neurons of the cortical plate, but not in the mature cortex (Francis et al. 1999). In adult mice, DCX expression indicates the presence of neurogenesis (e.g., Jessberger et al. 2005; Couillard-Despres et al. 2005). DCX-knockout mice provided the surprising information that DCX deficiency had no visible consequences for cortical development in this species, although DCX was required for normal lamination of the hippocampus (Corbo et al. 2002). By contrast, acute inactivation of DCX by in utero electroporation of RNAi produced disruptions of radial migration and formation of ectopic neurons in a subcortical band reminiscent of the dramatic phenotype of human SBH (Bai et al. 2003). To examine the possible compensatory activity of DCLK in DCX-knockout mice, mice mutant for both DCX and DCLK were generated (Koizumi et al. 2006; Deuel et al. 2006). Double knockout mice displayed profound cortical disorganization and axonal defects in fiber tracts such as the corpus callosum, anterior commissure, and internal capsule, suggesting that DCX and DCLK play synergistic roles in neuronal migration and axonal growth.

9.4.2
DCX and LIS1 Expression in Human Cortex

As described above, DCX and LIS1 are involved in cortical development through different functional activities, although mutations in both genes give rise to a similar human cortical phenotype characterized by decreased folding and abnormal lamination. We therefore asked whether the two proteins were expressed in the same cells, or alternatively, displayed distinct and characteristic expression patterns. To this end, we examined the expression of LIS1 and DCX from early embryonic stages to term (Meyer et al. 2002b).

The distribution of LIS1 appears quite generalized, because the protein is detected in virtually all cells of the developing telencephalon. This is not surprising if we take into account the variety of activities, summarized above, in which LIS1 has been implicated. In keeping with its ubiquitous distribution, LIS1 is probably expressed by progenitor cells as well as by migrating and differentiating neurons. Although we observed a particularly high signal in the SVZ and in Cajal–Retzius cells (Clark et al. 1997), the generalized expression pattern makes it difficult to infer more specific roles in cortical development and leads to the conclusion that all cortical cells might potentially be affected by LIS1 deficiency.

DCX expression is quite different. From very early stages we detected strong immunoreactivity in distinct neuronal populations along with negativity in well-defined cell compartments. In general, the ventricular zone was DCX negative, at least the proliferating and precursor populations (Fig. 25C). On the other hand, a few presumably postmitotic cells within the VZ compartment expressed high levels of DCX. Most strikingly, in the embryonic frontal lobe (Fig. 25C, D, E) DCX-positive cells were arranged in radial clusters spanning the whole width of the DCX-negative neuroepithelium.

We want to point out several remarkable aspects of these clusters: first, their spatial restriction to the rostral aspect of the pallium, the future frontal lobe. This early region-specific expression may be related to the characteristic severity of the X-linked lissencephalic/SBH syndrome in the frontal lobe (Pilz et al. 1998; Gleeson et al. 2000). Second, the early time point of the clusters' formation corresponds to an initial preplate stage, when the only known preplate neurons are the CR cells. In fact, Reelin immunostaining on parallel sections revealed that at least some cells within the clusters were also Reelin positive (compare Fig. 25B and D), suggesting a local and focal origin of a subset of CR cells. When we contrasted these findings with the expression of p73, a marker of CR cells derived from the cortical hem, we found no coincidence with the DCX-positive clusters (Meyer et al. 2002a, b), suggesting a heterogeneous origin of preplate CR cells. A third interesting aspect of the radial cell clusters is the possibility that they may be human-specific, or at least primate-specific, since comparable expression patterns have not been reported in mice.

The complexity of DCX expression increased at CS 18, when the SVZ appeared in the lateral aspect of the pallium, close to the cortico-striatal angle. An intense DCX-positive cell and fiber staining accompanied the emergence of the SVZ (Fig. 26) and its latero-medial progression during the following stages. In the advanced preplate at CS 20, apparently all neurons and many horizontal fibers expressed DCX, in contrast to the negativity of the proliferative VZ. However, single DCX-positive cells and fibers with large growth cones were occasionally observed in the VZ.

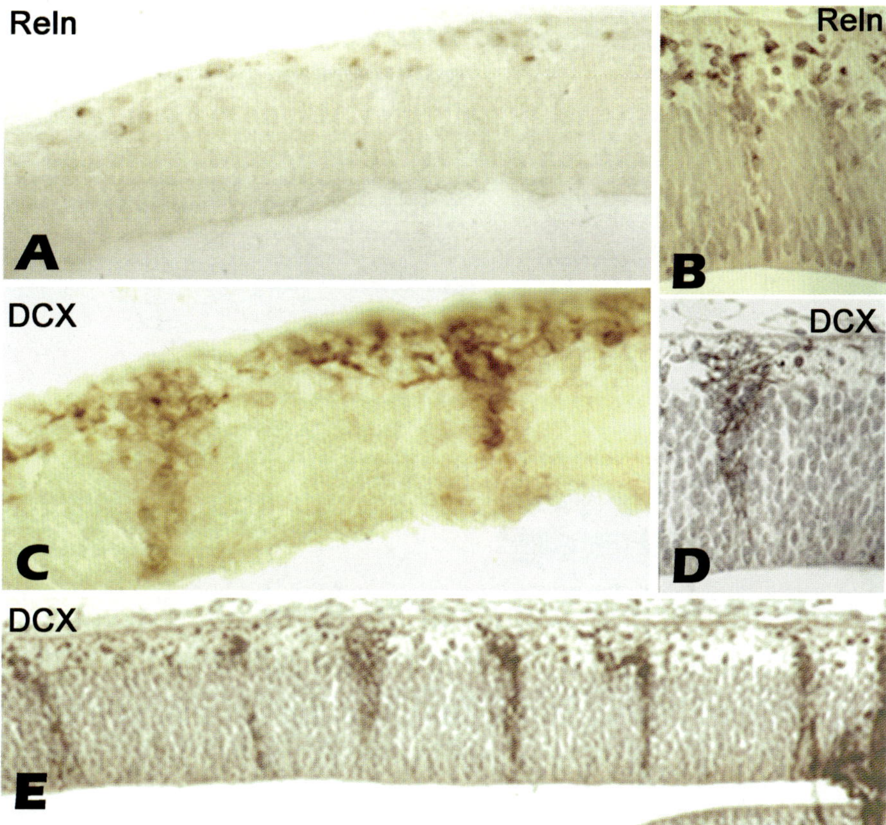

Fig. 25A–E Reelin and DCX in early corticogenesis. **A** Reelin (*Reln*) at 5.5 GW (CS 16) is mostly in the marginal layer. **C** DCX at CS 16 is expressed in radial columns which seem to supply neurons for the marginal layer. **B, D** At 6.5 GW (CS 19), showing radial columns of both **B** Reelin-positive and **D** DCX-positive cells in the future frontal lobe, suggesting that some early CR cells have a local origin. **E** CS 19, showing the conspicuous regular intervals of the radial DCX-positive columns

The appearance of the pioneer plate at 8 GW added a new degree of complexity in DCX expression, because the large neurons of the pioneer plate were DCX positive (Fig. 26). In turn, the neurons of the cortical plate expressed detectable levels of DCX only along their apical processes close to the MZ (Fig. 26). It is tempting to suggest that the initial condensation of the pioneer plate requires higher levels of DCX than the maintenance of the cortical plate, once the cells have detached from the radial glia fibers.

The SVZ increases in size during evolution, and DCX expression increases in parallel (Fig. 27). In our material, from approximately 12 GW onward, the number of DCX-positive neurons in the SVZ steadily increased to reach maximum values around 27–29 GW and then decreased by the end of gestation. Morphologically,

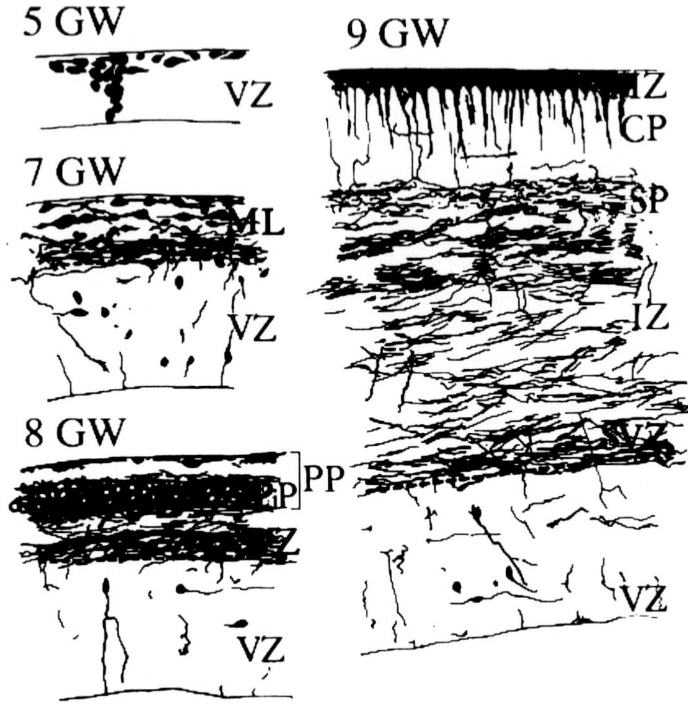

Fig. 26 Camera lucida drawings summarizing the expression pattern of DCX at different preplate stages and in the initial cortical plate stage. DCX is first expressed in radially oriented columns extending through the VZ. DCX is then present in virtually all cell populations that appear successively in the preplate (*PP*), as well as in the pioneer plate (*PiP*) at 8 GW. DCX also labels a fiber plexus in the SVZ that becomes more prominent during later stages. The staining pattern changes after the emergence of the cortical plate (*CP*) at 9 GW, when DCX is expressed in radial processes in the upper CP, as well as in non-radial fibers and neurons in the lower compartments of the cortical wall. (From Meyer et al. 2004, with permission)

DCX-ir neurons were small and horizontally oriented, in keeping with tangential migration. The co-expression of DCX and calretinin in many of these cells (Fig. 10) suggested a non-pyramidal phenotype. DCX is often used as a marker of newborn neurons (e.g., Jessberger et al. 2005; Couillard-Despres et al. 2005), and the proximity of DCX-expressing neurons to proliferating cells in the SVZ, along with a prominent clustering, suggested that neurons born in the SVZ began to express DCX while they migrated tangentially. In humans, GABAergic interneurons may be generated in the pallial SVZ (Letinic et al. 2002), which might explain the prominence of DCX in this compartment. On the other hand, recent findings in mice showed that radially migrating long-projecting neurons also rest in the SVZ and can travel tangentially before assuming a radial migratory pathway (Kriegstein and Noctor 2004). DCX may thus be involved in tangential migratory phases that may be common to pyramidal and non-pyramidal neurons.

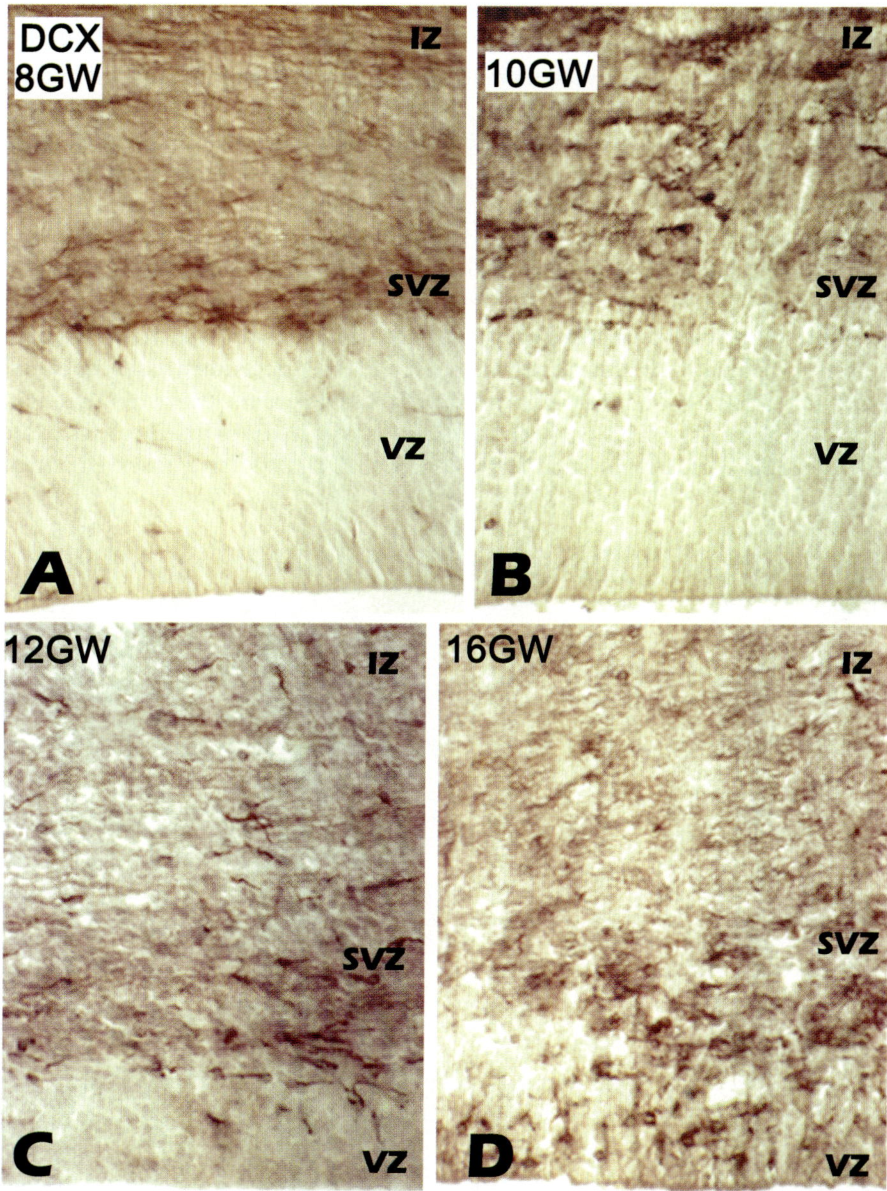

Fig. 27A–D DCX in the VZ and SVZ. **A** At 8 GW, the VZ is almost devoid of DCX-positive elements, while the SVZ contains an intensely DCX-stained fiber plexus and horizontally oriented neurons. **B** At 10 GW, the VZ is basically unchanged, but the number of horizontal cells increases in the VZ. **C** At 12 GW, the VZ is reduced in width, and horizontally oriented DCX-positive neurons in the SVZ seem to invade the VZ. **D** At 16 GW, the VZ is further reduced, and the number of horizontal DCX-positive neurons continues to increase, often forming large clusters (compare with Fig. 10)

The preference of DCX for non-radially oriented neurons was particularly striking in the intermediate zone from 12 to 18 GW. The comparison of DCX and LIS1 immunostaining clearly showed that radially migrating cells were grouped in clusters and expressed LIS1 but not DCX (Fig. 28). These data suggest that DCX is more closely associated with tangential migration, whereas LIS1 is universally expressed by radially and tangentially migrating neurons. Although mutations of both genes give rise to a similar lissencephalic phenotype, recent studies indicate that there may be differences in the arrangement of cortical neurons as a result of a preferential disruption of radial or tangential migration modes (Pancoast et al. 2005; Viot et al. 2004).

Our current analysis of genetically identified lissencephalic fetal brains is aimed at detecting possible differences between LIS1 and DCX-dependent lissencephalies, which may shed light on the pathogenesis of these disorders (Meyer and Francis, in preparation). In keeping with recent reports of a decrease of interneurons in the lissencephalic cortex (Viot et al. 2004; Pancoast et al. 2004), we observed a near

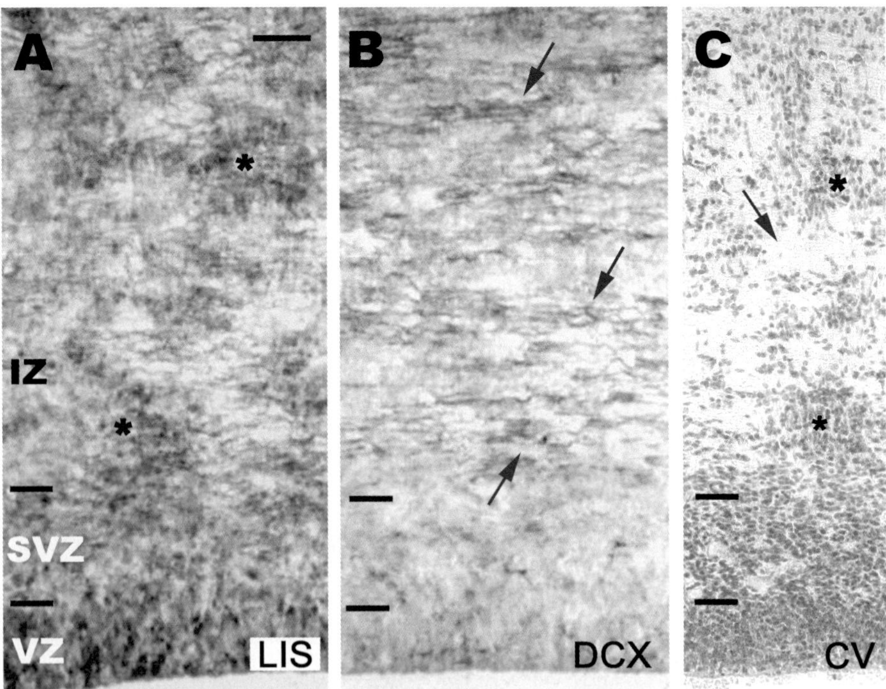

Fig. 28A–C DCX and LIS1 in normal fetal cortex at 15 GW. **A** LIS1 is expressed in proliferative and migratory compartments and marks groups of radially oriented cells (*asterisks*) that travel though the IZ. **B** By contrast, DCX is highly expressed in non-radially oriented cells in VZ, SVZ, and IZ, as well as in horizontal fibers in the IZ (*arrows*). **C** Cresyl violet (*CV*) shows radial cell clusters in IZ (*asterisks*) and complementary, unstained fiber compartments (*arrows*). *Scale bar*: 100 μm. (From Francis et al. 2006, with permission)

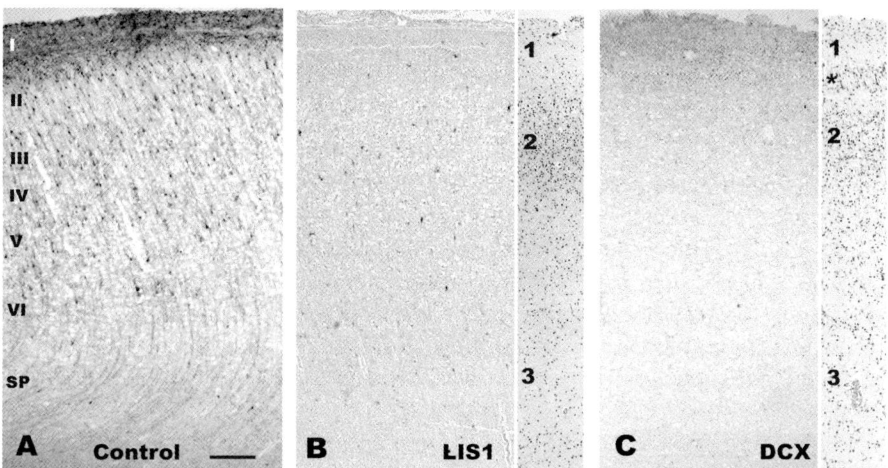

Fig. 29A–C Calretinin expression in LIS1- and DCX-mutated brains. Calretinin-positive interneurons in the frontal cortex of **A** a normal newborn, 40 GW; **B** a LIS1-mutation, 34 GW; and **C** a DCX mutation, 35 GW. **B** and **C** are accompanied by their corresponding *cresyl violet* control sections, to indicate the position of abnormal layers 1–3, described for four-layered type 1 lissencephaly (Jellinger and Rett 1976). These layers do not correspond to the six layers of the normal neocortex, which are indicated in **A**. The *asterisk* in C indicates a narrow cell layer characteristic of mutated DCX lissencephalies that we have not found in LIS1-mutated cases. In the lissencephalic cases, the number of calretinin-positive interneurons is dramatically reduced (**B**, or even almost abolished **C**. *Scale bar*: 200 µm.) (From Francis et al. 2006, with permission)

absence of calretinin and calbindin-positive neurons in the upper cell layers of both LIS1 and DCX mutations (Fig. 29; Francis et al. 2006). Further studies will hopefully unravel the mechanistic basis of LIS and DCX-dependent lissencephalies.

10
ARX Deficiency: A Novel Type of Lissencephaly

There are several other known gene mutations which cause type1 lissencephaly. X-linked lissencephaly with absent corpus callosum and ambiguous genitalia is a severe malformation syndrome that is caused by mutations of the X-linked aristaless-related homeobox gene *(ARX)* and has been observed only in boys (Kitamura et al. 2002; Hartmann et al. 2004). MRI studies showed anterior pachygyria and posterior agyria with a mildly thickened cortex, agenesis of the corpus callosum, and dysplastic basal ganglia. Associated brain malformations included absence of the corpus callosum, cavitated basal ganglia and microcephaly Neuropathology examination revealed abnormal cortical lamination with a predominance of pyramidal neurons, suggesting a disruption of both radial and tangential migration, as well as the presence of heterotopic neurons in the white matter (Bonneau et al. 2002). Clinically, the syndrome is accompanied by neonatal epilepsy, hypothalamic dysfunction, chronic diarrhea, and ambiguous genitalia and leads commonly to early death (Kato et al. 2004). *ARX* is also the causative gene of X-linked infantile spasms or West syndrome and of certain cases of X-linked syndromic and non-syndromic mental retardation, found in patients with dystonia and patients with myoclonic epilepsy and spasticity (Bienvenu et al. 2002; Stromme et al. 2002).

The *ARX*-predicted protein (ARX) belongs to the largest classes of homeoproteins, the paired (*Prd*) class. In addition to its paired/Q50 central homeodomain, ARX is characterized by a 14 amino acid C-terminal aristaless domain and by an octapeptide domain near the N-terminus, designated as the goosecoid engrailed homology or the eh-1 domain in the engrailed (En) homeoprotein (Mailhos et al. 1998).

The expression of Arx in the mouse forebrain includes Dlx-expressing territories, such as the ventral thalamus, parts of the hypothalamus, and the ganglionic eminences and their derivatives in the subpallial telencephalon; like the Dlx genes, it is expressed in cortical GABAergic neurons (Cobos et al. 2005; Poirier et al. 2004). The striking epileptogenicity of X-linked lissencephaly with abnormal genitalia and West's syndrome associated with *ARX* mutations is thought to be caused by a disorder of interneurons implicating a tangential migration defect, and the term "interneuronopathy" has been proposed (Kato and Dobyns 2005). Certainly, the ARX-related lissencephaly may not be the only "interneuronopathy," and it remains to be seen whether DCX lissencephaly has similar features although induced by a different molecular pathway.

11
Final Considerations: The Unique Features of Human Brain Development

In this monograph, we have extensively described the specific features that define human cortical development, and we have established comparisons with what is known on cortical development in rodents. In this final paragraph, we summarize the main differences we have observed at a morphological level between rodent and human cortical development. First, we want to point out the comparatively long duration of the initial phase of corticogenesis, when the only differentiated cell class outside the proliferating ventricular zone are Reelin/p73-expressing CR cells. This protracted period may be related with the establishment of a large pool of primary progenitor cells in the VZ as a first requirement for building a large cortex. A second important difference is the complex cell composition of the human preplate, which includes a variety of tangentially migrating cell populations, such as Cajal–Retzius cells migrating from the cortical hem in a medial to lateral direction, and GABAergic "monolayer cells" possibly arriving from the ganglionic eminences. A third important difference is the condensation of a "pioneer plate" below the preplate populations and its subsequent splitting into superficial and deep pioneer cells by newly arriving cortical plate cells. This developmental step has not been described in other species, perhaps because only the human brain provides the sufficient temporo-spatial resolution to allow recognition.

The next phases of development have been addressed in monkeys and human, in which the subplate compartment undergoes an extraordinary enlargement and differentiation through the continuous addition of new cells. A similar phenomenon takes place in the MZ, where CR cells still increase in number after the establishment of the cortical plate. The structural complexity of human CR cells and the density of their axonal fiber plexus go far beyond those of their rodent counterparts. Furthermore, the subpial granular layer is a distinctive feature of the fetal human cortex, and we propose that it may be related to the tangential migration of interneurons from the ganglionic eminences via the anterior perforated substance into the cortical MZ. We have addressed the possibility that certain types of interneurons are born in the SVZ, another important divergence from the rodent model. Examples of gene mutations and their severe consequences on human brain development have been discussed for the *REELIN*, *LIS1*, and *DCX* genes. In all these cases, the consequences of gene mutations are much more dramatic in human than in the mouse, and it is thus necessary to refine our knowledge on the cellular and molecular mechanisms that regulate neuronal migration in human cortical development.

A final point is the wealth of data regarding the acquisition of regional identities through the graded expression of genes or gene families in specific domains of the rodent telencephalon, a theme that has become a major issue during the last few years. Thus far, there are only a few studies that address cortical regionalization in the human cortex, but it may be predicted that this process is much more complex than in rodents, taking into account the most striking macroscopic features in addition to the size and convoluted surface: the major division of the human cortex into lobes, the distinctive organization of the ventricular system with the presence of temporal and occipital horns, and the exclusive localization of the hippocampus in the temporal lobe. It will be a formidable task to understand the developmental mechanisms that lead to the unique complexity of the human cortex.

References

Abraham H, Meyer G (2003) Reelin-expressing neurons in the postnatal and adult human hippocampal formation. Hippocampus 13:715–727

Abraham H, Pérez-García CG, Meyer G (2004) p73 and Reelin in Cajal-Retzius cells of the developing human hippocampal formation. Cerebral Cortex 14:484–495

Abu-Khalil A, Fu L, Grove EA, Zecevic N, Geschwind DH (2004) Wnt genes define distinct boundaries in the developing human brain: Implications for human forebrain patterning. J Comp Neurol 474:276–288

Aicardi J (1991) The agyria-pachygyria complex: a spectrum of cortical malformations. Brain Dev, 13, 1–8

Akbarian S, Kim JJ, Potkin SG, Hetrick WP, Bunney WE Jr, Jones EG (1996) Maldistribution of interstitial neurons in prefrontal white matter of the brains of schizophrenic patients. Arch Gen Psychiatry 53:425–436

Alcantara S, Ruiz M, D'Arcangelo G, Ezan F, de Lecea L, Curran T, Sotelo C, Soriano E (1998) Regional and cellular patterns of Reelin mRNA expression in the forebrain of the developing and adult mouse. J Neurosci 18:7779–7799

Allendoerfer KL, Schatz CJ (1994) The subplate, a transient neocortical structure: its role in the development of connections between thalamus and cortex. Annu Rev Neurosci 17:185–218

Altman J, Bayer SA (1991) Neocortical development. Raven Press New York

Álvarez-Buylla A, Garcia-Verdugo JM (2002) Neurogenesis in adult ventricular zone. J Neurosci 22:629–634

Anderson SA, Eisenstat DD, Shi L, Rubenstein JL (1997) Interneuron migration from basal forebrain to neocortex: dependence on Dlx genes. Science 278:474–476

Anderson SA, Kaznowski CE, Horn C, Rubenstein JL, McConnell SK (2002) Distinct origins of neocortical projection neurons and interneurons in vivo. Cereb Cortex 12:702–709

Anderson SA, Marin O, Horn C, Jennings K, Rubenstein JL (2001) Distinct cortical migrations from the medial and lateral ganglionic eminences. Development 128:353–363

Anderson SA, Mione M, Yun K, Rubenstein JL (1999) Differential origins of neocortical projection and local circuit neurons: role of Dlx genes in neocortical interneuronogenesis. Cereb Cortex 9:646–654

Anderson SA, Volk DW, Lewis DA (1996) Increased density of microtubule associated protein 2-immunoreactive neurons in the prefrontal white matter of schizophrenic subjects. Schizophr Res 19:111–119

Ang ESBC, Haydar TF, Glunic V, Pakic P (2003) Four-dimensional migratory coordinates of GABAergic interneurons in the developing mouse cortex. J Neurosci 23:5805–5815

Angevine JB, Sidman RL (1961) Autoradiographic study of cell migration during histogenesis of cerebral cortex in the mouse. Nature 192:766–768

Anton ES, Kreidberg JA, Rakic P (1999) Distinct functions of alpha3 and alpha(v) integrin receptors in neuronal migration and laminar organization of the cerebral cortex. Neuron 22:277–289

Arnaud L, Ballif BA, Forster E, Cooper JA (2003) Fyn tyrosine kinase is a critical regulator of disabled-1 during brain development. Curr Biol 13:9–17

Assadi AH, Zhang G, Beffert U, McNeil RS, Renfro AL, Niu S, Quattrocchi CC, Antalffy BA, Sheldon M, Armstrong DD, Wynshaw-Boris A, Herz J, D'Arcangelo G, Clark GD (2003) Interaction of Reelin signaling and Lis1 in brain development. Nat Genet 35:270–276

Aumais JP, Tunstead JR, McNeil RS, Schaar BT, McConnell SK, Lin SH, Clark GD, Yu-Lee LY (2001) NudC associates with Liss1 and dynein motor at the leading pole of neurons. J Neurosci 21:RC187

Bai J, Ramos RL, Ackman JB, Thomas AM, Lee RV, LoTurco JJ (2003) RNAi reveals doublecortin is required for radial migration in rat neocortex. Nat Neurosci 6:1277–1283

Ballif BA, Arnaud L, Arthur WT, Guris D, Imamoto A, Cooper JA (2004) Activation of a Dab1/CrkL/C3G/Rap1 pathway in Reelin-stimulated neurons. Curr Biol 14:606–610

Bar I, Lambert de Rouvroit C, Goffinet AM (2000) The evolution of cortical development. An hypothesis based on the role of the Reelin signaling pathway. Trends Neurosci 23:633–638

Bar I, Tissir F, Lambert de Rouvroit C, De Backer O, Goffinet AM (2003) The gene encoding disabled-1 (DAB1), the intracellular adaptor of the Reelin pathway, reveals unusual complexity in human and mouse. J Biol Chem 278:5802–5812

Barkovich, AJ, Kuzniecky, RI, Jackson, GD, Guerrini, R, Dobyns, WB (2001) Classification system for malformations of cortical development: update 2001. Neurology, 57:2168–78

Barkovich AJ, Kuziecky RI (2000) Gray matter heterotopia. Neurology 55:1603–1608

Bayer SA, Altman J (1990) Development of layer I and the subplate in the rat neocortex. Exp Neurol 107:48–62

Bayer SA, Altman J (2004) The human brain during the third trimester. CRC Press, Boca Raton

Beasley CL, Cotter DR, Everal IP (2002) Density and distribution of white matter neurons in schizophrenia, bipolar disorder and major depressive disorder: no evidence for abnormalities of neuronal migration. Mol Psychiatry 7:564–570

Beffert U, Weeber EJ, Durudas A, Qiu S, Masiulis I, Sweatt JD, Li WP, Adelmann G, Frotscher M, Hammer RE, Herz J (2005) Modulation of synaptic plasticity and memory by Reeling involves differential splicing of the lipoprotein receptor ApoER2. Neuron 47:567–579

Beffert U, Weeber EJ, Morfini G, Ko J, Brady ST, Tsai LH, Sweatt JD, Herz J (2004) Reelin and cyclin-dependent kinase 5-dependent signals cooperate in regulating neuronal migration and synaptic transmission. J Neurosci 24:1897–1906

Belichenko PV, Vogt Weisenhorn DM, Myklossy J, Celio MR (1995) Calretinin-positive Cajal-Retzius cells persist in the adult human neocortex. NeuroReport 6:1869–1874

Bellion A, Baudoin JP, Alvarez C, Bornens M, Metin C (2005) Nucleokinesis in tangentially migrating neurons comprises two alternating phases: Forward migration of the Golgi/centrosome associated with centrosome splitting and myosin contraction at the rear. J Neurosci 25:5691–5699

Benhayon D, Magdaleno S, Curran T (2003) Binding of purified Reelin to ApoER2 and VLDLR mediates tyrosine phosphorylation of Disabled-1. Brain Res Mol Brain Res 112:33–45

Bernasconi A, Martinez V, Rosa-Neto P, D'Agostino D, Bernasconi N, Berkovic S, MacKay M, Harvey AS, Palmini A, da Costa JC, Paglioli E, Kim HI, Connolly M, Olivier A, Dubeau F, Andermann E, Guerrini R, Whisler W, de Toledo-Morrell L, Morrell F, Andermann F (2001) Surgical resection for intractable epilepsy in "double cortex" syndrome yields inadequate results. Epilepsia 42:1124–1129

Bernier B, Bar I, Pieau C, Lambert De Rouvroit C, Goffinet AM (1999) Reelin mRNA expression during embryonic brain development in the turtle Emys orbicularis. J Comp Neurol 413:463–479

Bernier B, Bar I, D'Arcangelo G, Curran T, Goffinet AM (2000) Reelin mRNA expression during embryonic brain development in the chick. J Comp Neurol 422:448–463

Berry M, Rogers AW (1965) The migration of neuroblasts in the developing cerebral cortex. J Anat 99:691–709

Bielle F, Griveau A, Narboux-Neme N, Vigneau S, Sigrist M, Arber S, Wassef M, Pierani A (2005) Multiple origins of Cajal-Retzius cells at the borders of the developing pallium. Nat Neurosci 8:1002–1012

Bienvenu T, Poirier K, Friocourt G, Bahi N, Beaumont D, Fauchereau F, Ben Jeema L, Zemni R, Vinet MC, Francis F, Couvert P, Gomot M, Moraine C, van Bokhoven H, Kalscheuer V, Frints S, Gecz J, Ohzaki K, Chaabouni H, Fryns JP, Desportes V, Beldjord C, Chelly J (2002) ARX, a novel Prd-class-homeobox gene highly expressed in the telencephalon, is mutated in X-linked mental retardation. Hum Mol Genet 11:981–991

Bishop KM, Goudreau G, O'Leary DD (2000) Regulation of area identity in the mammalian neocortex by Emx2 and Pax6. Science 288:344–349

Bishop KM, Rubenstein JL, O'Leary DD (2002) Distinct actions of Emx1, Emx2, and Pax6 in regulating the specification of areas in the developing neocortex. J Neurosci 22:7627–7638

Bishop KM, Garel S, Nakagawa Y, Rubenstein JL, O'Leary DD (2003) Emx1 and Emx2 cooperate to regulate cortical size, lamination, neuronal differentiation, development of cortical efferents, and thalamocortical pathfinding. J Comp Neurol 457:345–360

Bittman K, Owens DF, Kriegstein AR, Lo Turco JJ (1997) Cell coupling and uncoupling in the ventricular zone of developing neocortex. J Neurosci 17:7037–7044

Bock HH, Herz J (2003) Reelin activates SRC family tyrosine kinases in neurons. Curr Biol 13:18–26

Bock HH, Jossin Y, Liu P, Forster E, May P, Goffinet AM, Herz J (2003) Phosphatidylinositol 3-kinase interacts with the adaptor protein Dab1 in response to Reelin signaling and is required for normal cortical lamination. J Biol Chem 278:38772–38779

Bond J, Roberts E, Mochida GH, Hampshire DJ, Scott S, Askham JM, Springell K, Mahadevan M, Crow YJ, Markham AF, Walsh CA, Woods CG (2002) ASPM is a major determinant of cerebral cortical size. Nat Genet 32:316–320

Bond J, Roberts E, Springell K, Lizarraga SB, Scott S, Higgins J, Hampshire DJ, Morrison EE, Leal GF, Silva EO, Costa SM, Baralle D, Raponi M, Karbani G, Rashid Y, Jafri H, Bennett C, Corry P, Walsh CA, Woods CG (2005) A centrosomal mechanism involving CDK5RAP2 and CENPJ controls brain size. Nat Genet 37:353–355

Bonneau D, Toutain A, Laquerriere A, Marret S, Saugier-Veber P, Barthez MA, Radi S, Biran-Mucignat V, Rodriguez D, Gelot A.(2002) X-linked lissencephaly with absent corpus callosum and ambiguous genitalia (XLAG): clinical, magnetic resonance imaging, and neuropathological findings. Ann Neurol 51:340–349

Borrell V, Del Rio JA, Alcantara S, Derer M, Martinez A, D'Arcangelo G, Nakajima K, Mikoshiba K, Derer P, Curran T, Soriano E (1999) Reelin regulates the development and synaptogenesis of the layer-specific entorhino-hippocampal connections. J Neurosci 19:1345–1358

Boulder Committee Report (1970) Embryonic vertebrate system: Revised terminology. Anat Rec 166:257–262

Boycott KM, Flavelle S, Bureau A, Glass HC, Fujiwara TM, Wirrell E, Davey K, Chudley AE, Scott JN, McLeod DR, Parboosingh JS (2005) Homozygous deletion of the very low density lipoprotein receptor gene causes autosomal recessive cerebellar hypoplasia with cerebral gyral simplification. Am J Hum Genet 77:477–83

Broca P (1878) Anatomie comparee des circonvolutions cerebrales. Le grand lobe limbique et la scissure limbique dans la serie des mammiferes. Rev Anthropol I, Ser 8:385–498

Brodmann K (1909) Vergleichende Lokalisationslehre der Grosshirnrinde in ihren Prinzipien dargestellt auf Grund des Zellenbaues, Barth, Leipzig

Brown MT, Cooper JA (1996) Regulation, substrates and functions of src. Biochim Biophys Acta 1287:121–149

Brun A (1965) The subpial granular layer of the fetal cerebral cortex in man. Its ontogeny and significance in congenital cortical malformations. Acta Pathol Microbiol Scand Suppl 179:3–98

Bulfone P, di Blas E, Gulisano MH, Frohmann MA, Martin GR, Rubenstein JLR (1993) Spatially restricted expression of Dlx-1, Dlx-2 (Tes-1), Gbx-2 and Wnt-3 in the embryonic day 12.5 mouse forebrain defines potential transverse and longitudinal segmental boundaries. J Neurosci 13:3155–3172

Burgess HA, Martinez S, Reiner O (1999) KIAA0369, doublecortin-like kinase, is expressed during brain development. J Neurosci Res 58:567–575

Burgess HA, Reiner O (2000) Doublecortin-like kinase is associated with microtubules in neuronal growth cones. Mol Cell Neurosci 16:529–541

Burkhalter A (1993) Development of forward and feedback connections between areas V1 and V2 of human visual cortex. Cereb Cortex 3:476–487

Burkhalter A, Bernardo KL, Charles V (1993) Development of local circuits in human visual cortex. J Neurosci 13:1916–1931

Cáceres M, Lachuer J, Zapala MA, Redmond JC, Kudo L, Geschwind DH, Lockhart DJ, Preuss TM, Barlow C (2003) Elevated gene expression levels distinguish human from non-human. Proc Natl Acad Sci USA 100:13030–13035

Cahana A, Escamez T, Nowakowski RS, Hayes NL, Giacobini M, von Holst A, Shmueli O, Sapir T, McConnell SK, Wurst W, Martinez S, Reiner O (2001) Targeted mutagenesis of Lis1 disrupts cortical development and LIS1 homodimerization. Proc Natl Acad Sci U S A 98:6429–6434

Carroll SB (2003) Genetics and the making of Homo sapiens. Nature 422:849–857

Caspi M, Atlas R, Kantor A, Sapir T, Reiner O (2000) Interaction between LIS1 and doublecortin, two lissencephaly gene products. Hum Mol Genet 9:2205–2213

Caviness VS Jr (1976) Patterns of cell and fiber distribution in the neocortex of the reeler mutant mouse. J Comp Neurol 170:435–448

Caviness VS Jr (1982) Neocortical histogenesis in normal and reeler mice: a developmental study based upon [3H]thymidine autoradiography. Brain Res 256:293–302

Caviness VS Jr, Sidman RL (1973) Time of origin of corresponding cell classes in the cerebral cortex of normal and reeler mutant mice: an autoradiographic analysis. J Comp Neurol 148:141–151

Caviness VS Jr, Takahashi T, Nowakowski RS (1995) Numbers, time and neocortical neurogenesis: a general developmental and evolutionary model. Trends Neurosci 18:379–383

Cecchi C, Boncinelli E (2000) Emx homeogenes and mouse brain development. Trends Neurosci 23:347–352

Chae T, Kwon YT, Bronson R, Dikkes P, Li E, Tsai LH (1997) Mice lacking p35, a neuronal specific activator of Cdk5, display cortical lamination defects, seizures, and adult lethality. Neuron 18:29–42

Chan CH, Godinho LN, Thomaidou D, Tan SS, Gulisano M, Parnavelas JG (2001) Emx1 is a marker for pyramidal neurons of the cerebral cortex. Cereb Cortex 11:1191–1198

Chang BS, Walsh CA (2003) Mapping form and function in the human brain: The emerging field of functional neuroimaging in cortical malformations. Epil Behavior 4:618–625

Chen Y, Beffert U, Ertunc M, Tang TS, Kavalali ET, Bezprozvanny I, Herz J (2005) Reelin modulates NMDA receptor activity in cortical neurons. J Neurosci 25:8209–8216

Chi JG, Dooling EC, Gilles FH (1977) Gyral development of the human brain. Ann Neurol 1:86–93

Chong SS, Pack SD, Roschke AV, Tanigami A, Carrozzo R, Smith AC, Dobyns WB, Ledbetter DH (1997) A revision of the lissencephaly and Miller-Dieker syndrome critical regions in chromosome 17p13.3. Hum Mol Genet 6:147–155

Chun JJM, Shatz CJ (1989a) The earliest-generated neurons of the cat cerebral cortex: characterization by MAP2 and neurotransmitter immunochemistry during fetal life. J Neurosci 9:1648–1667

Chun JJM, Shatz CJ (1989b) Interstitial cells of the adult neocortical white matter are the remnant of the early generated subplate neuron population. J Comp Neurol 282:555–569

Clark GD (2004) The classification of cortical dysplasias through molecular genetics. Brain Dev 26:351–362

Clark GD, Mizuguchi M, Antalffy B, Barnes J, Armstrong D (1997) Predominant localization of the LIS family of gene products to Cajal-Retzius cells and ventricular neuroepithelium in the developing human cortex. J Neuropathol Exp Neurol 56:1044–1052

Cobos I, Broccoli V, Rubenstein JL (2005) The vertebrate ortholog of Aristaless is regulated by Dlx genes in the developing forebrain. J Comp Neurol 483:292–303

Colombo E, Galli R, Cossu G, Gecz J, Broccoli V (2004) Mouse orthologue of ARX, a gene mutated in several X-linked forms of mental retardation and epilepsy, is a marker of adult neural stem cells and forebrain GABAergic neurons. Dev Dyn 231:631–9

Corbin JG, Nery S, Fishell G (2001) Telencephalic cells take a tangent: non-radial migration in the mammalian forebrain. Nat Neurosci 4:1177–1182

Corbo JC, Deuel TA, Long JM, LaPorte P, Tsai E, Wynshaw-Boris A, Walsh CA (2002) Doublecortin is required in mice for lamination of the hippocampus but not the neocortex. J Neurosci 22:7548–7557

Costagli A, Kapsimali M, Wilson SW, Mione M (2002) Conserved and divergent patterns of Reelin expression in the zebrafish central nervous system. J Comp Neurol. 450:73–93

Couillard-Despres S, Winner B, Schaubeck S, Aigner R, Vroemen M, Weidner N, Bogdahn U, Winkler J, Kuhn HG, Aigner L (2005) Doublecortin expression levels in adult brain reflect neurogenesis. Eur J Neurosci 21:1–14

Couillard-Despres S, Winner B, Schaubeck S, Aigner R, Vroemen M, Weidner N, Bodgan U, Winkler J, Kuhn HG, Aigner L (2005) Doublecortin expression levels in adult brain reflect neurogenesis. Europ J Neurosci 21:1–14

D'Arcangelo G (2005) ApoER2: a reelin receptor to remember. Neuron 47:471–473

D'Arcangelo G, Miao GG, Chen SC, Soares HD, Morgan JI, Curran T (1995) A protein related to extracellular matrix proteins deleted in the mouse mutant reeler. Nature 374:719–723

D'Arcangelo G, Nakajima K, Miyata T, Ogawa M, Mikoshiba K, Curran T (1997) Reelin is a secreted glycoprotein recognized by the CR-50 monoclonal antibody J Neurosci 17:23–31

D'Arcangelo G, Homayouni R, Keshvara L, Rice DS, Sheldon M, Curran T (1999) Reelin is a ligand for lipoprotein receptors. Neuron 24:471–479 de Bergeyck V, Naerhuyzen B, Goffinet AM, Lambert de Rouvroit C (1998) A panel of monoclonal antibodies against Reelin, the extracellular matrix protein defective in reeler mutant mice. J Neurosci Methods 82:17–24

de Bergeyck V, Nakajima K, Lambert de Rouvroit C, Naerhuyzen B, Goffinet AM, Miyata AT, Ogawa M, Mikoshiba K (1997) A truncated Reelin protein is produced but not secreted in the "Orleans" reeler mutation (Reln[rl-Orl]). Mol Brain Res 50:85–90

De Carlos JA, O'Leary DDM (1992) Growth and targeting of subplate axons and establishment of major cortical pathways. J Neurosci 12:1194–1211

De Carlos JA, Lopez-Mascaraque L, Valverde F (1996) Dynamics of cell migration from the lateral ganglionic eminence in the rat. J Neurosci 16:6146–6156

DeDiego I, Smith-Fernandez A, Fairen A (1994) Cortical cells that migrate beyond area boundaries: characterization of an early neuronal population in the lower intermediate zone of prenatal rats. Eur J Neurosci. 6:983–997

Dehay C, Giroud P, Berland M, Smart I, Kennedy H (1993) Modulation of the cell cycle contributes to the parcellation of the primate visual cortex. Nature 366:464–466

Del Rio JA, Heimrich B, Borrell V, Förster E, Drakew A, Alcántara S, Nakajima K, Miyata T, Ogawas M, Mikoshiba K, Derer P, Frotscher M, Soriano E (1997) A role for Cajal-Retzius cells and reelin in the development of hippocampal connections. Nature 385:70–74

Del Rio JA, Martinez, Fonseca M, Auladell C, Soriano E (1995) Glutamate-like immunoreactivity and fate of Cajal-Retzius cells in the murine cortex as identified with calretinin antibody. Cereb Cortex 5:13–21

Denaxa M, Chan CH, Schachner M, Parnavelas JG, Karagogeos D (2001) The adhesion molecule TAG-1 mediates the migration of cortical interneurons from the ganglionic eminence along the corticofugal fiber system. Development 128:4635–4644

Derer P (1979) Evidence for the occurrence of early modifications in the 'glia limitans' layer of the neocortex of the reeler mutant mouse. Neurosci Lett 13:195–202

Derer P, Derer M (1990) Cajal-Retzius cell ontogenesis and death in mouse brain visualized with horseradish peroxidase and electron microscopy. Neuroscience 36:839–856

Derer P, Derer M, Goffinet A (2001) Axonal secretion of Reelin By Cajal-Retzius cells: Evidence from comparison of normal and $Reln^{Orl}$ mutant mice. J Comp Neurol 440:136–143

des Portes V, Pinard JM, Billuart P, Vinet MC, Koulakoff A, Carrie A, Gelot A, Dupuis E, Motte J, Berwald-Netter Y, Catala M, Kahn A, Beldjord C, Chelly J (1998a) A novel CNS gene required for neuronal migration and involved in X-linked subcortical laminar heterotopia and lissencephaly syndrome. Cell 92:51–61

des Portes V, Francis F, Pinard JM, Desguerre I, Moutard ML, Snoeck I, Meiners LC, Capron F, Cusmai R, Ricci S, Motte J, Echenne B, Ponsot G, Dulac O, Chelly J, Beldjord C (1998b) Doublecortin is the major gene causing X-linked subcortical laminar heterotopia (SCLH) Hum Mol Genet 7:1063–1070

De Silva U, D'Arcangelo G, Braden VV, Chen J, Miao GG, Curran T, Green ED (1997) The human reelin gene: isolation, sequencing, and mapping on chromosome 7. Genome Res 7:157–164

Deuel TA, Liu JS, Corbo JC, Yoo SY, Rorke-Adams LB, Walsh CA (2006) Genetic interactions between doublecortin and doublecortin-like kinase in neuronal migration and axon outgrowth. Neuron 49:41–53

Dobbing J, Sands J (1973) Quantitative growth and development of human brain. Arch Dis Child 48:757–767

Dobyns WB, Truwit CL (1995) Lissencephaly and other malformations of cortical development: 1995 update. Neuropediatrics 26:132–147

Dobyns WB, Reiner O, Carrozzo R, Ledbetter DH (1993) Lissencephaly: A human brain malformation associated with deletion of the LIS1 gene located at chromosome 17p13. JAMA 270:2838–2842

Dobyns WB, Andermann E, Andermann F, Czpansky-Beilman D, Dubeau F, Dulac O, Guerrini R, Hirsch B, Ledbetter DH, Nee NS, Motte J, Pinard JM, Radtke RA, Ross ME, Tampieri D, Walsh CA, Truwit CL (1996) X-linked malformations of neuronal migration. Neurology 47:331–339

Donoghue MJ, Rakic P (1999) Molecular gradients and compartments in the embryonic primate cerebral cortex. Cereb Cortex 9:586–600

Drakew A, Frotscher M, Deller T, Ogawa M, Heimrich B (1998) Developmental distribution of a reeler gene-related antigen in the rat hippocampal formation visualized by CR-50 immunocytochemistry. Neuroscience 82:1079–1086

Dulabon L, Olson EC, Taglienti MG, Eisenhuth S, McGrath B, Walsh CA, Kreidberg JA, Anton ES (2000) Reelin binds alpha3beta1 integrin and inhibits neuronal migration. Neuron 27:33–44

Eastwood SL, Harrison PJ (2003) Interstitial white matter neurons express less reelin and are abnormally distributed in schizophrenia: towards an integration of molecular and morphologic aspects of the neurodevelopmental hypothesis. Mol Psychiatry 8:821–831

Eastwood SL, Harrison PJ (2005) Interstitial white matter neuron density in the dorsolateral prefrontal cortex and parahippocampal gyrus in schizophrenia. Schizophr Res 79:181–188

Englund C, Fink A, Lau C, Pham D, Daza RAM, Bulfone A, Kowalczyk T, Hevner RF (2005) Pax6, Tbr2, and Tbr1 are expressed sequentially by radial glia, intermediate progenitor cells, and postmitotic neurons in developing neocortex. J Neurosci 25:247–251

Estivill-Torrus G, Pearson H, Van Heyningen V, Price DJ, Rashbass P (2002) Pax6 is required to regulate the cell cycle and the rate of progression from symmetrical to asymmetrical division in mammalian cortical progenitors. Development 129:455–466

Evans PD, Anderson JR, Vallender EJ, Gilbert SL, Malcom CM, Dorus S, Lahn BT (2004a) Adaptive evolution of ASPM, a major determinant of cerebral cortical size in humans. Hum Mol Genet 13:489–494

Evans PD, Anderson JR, Vallender EJ, Choi SS, Lahn BT (2004b) Reconstructing the evolutionary history of microcephalin, a gene controlling human brain size. Hum Mol Genet 13:1139–1145

Fairen A, Cobas A, Fonseca M (1986) Times of generation of glutamic acid decarboxylase immunoreactive neurons in mouse somatosensory cortex. J Comp Neurol. 251:67–83

Fairén A, de Felipe J, Regidor J (1984) Non pyramidal neurons: general account. In: Cerebral Cortex (editors Peters and Jones, New York), Vol 1:6

Fatemi SH, Earle JA, McMenomy T (2000) Reduction in Reelin immunoreactivity in hippocampus of subjects with schizophrenia, bipolar disorder and major depression. Mol Psychiatry 5:654–663

Faulkner NE, Dujardin DL, Tai CY, Vaughan KT, O'Connell CB, Wang Y, Vallee RB (2000) A role for the lissencephaly gene LIS1 in mitosis and cytoplasmic dynein function. Nat. Cell Biol 2:784–791

Fees-Higgin A, Larroche J-C (1987) Development of the human fetal brain. Masson Éditeur Paris

Feng Y, Olson EC, Stukenberg PT, Flanagan LA, Kirschner MW, Walsh CA (2000) LIS1 regulates CNS lamination by interacting with mNudE, a central component of the centrosome. Neuron 28:665–679

Fogli A, Guerrini R, Moro F, Fernandez-Alvarez E, Livet MO, Renieri A, Cioni M, Pilz DT, Veggiotti P, Rossi E, Ballabio A, Carrozzo R (1999) Intracellular levels of the LIS1 protein correlate with clinical and neuroradiological findings in patients with classical lissencephaly. Ann Neurol 45:154–161

Fonseca M, Del Rio JA, Martinez A, Gomez S, Soriano E (1995) Development of calretinin immunoreactivity in the rat. J Comp Neurol 361:177–192

Forman MS, Squier W, Dobyns WB, Golden JA (2005) Genotypically defined lissencephalies show distinct pathologies. J Neuropathol Exp Neurol 64:847–857

Francis F, Koulakoff A, Boucher D, Chafey P, Schaar B, Vinet MC, Friocourt G, McDonnell N, Reiner O, Kahn A, McConnell SK, Berwald-Netter Y, Denoulet P, Chelly J (1999) Doublecortin is a developmentally regulated, microtubule-associated protein expressed in migrating and differentiating neurons. Neuron 23:247–256

Francis F, Meyer G, Fallet C, Moreno S, Kappeler C, Cabrera Socorro A, Beldjord C, Chelly J (2006) Human disorders of cortical development – from past to present. Eur. J. Neurosci. 23:877–893

Friede RL (1989) Developmental Neuropathology. Springer, Berlin, Heidelberg New York

Friauf E, McConnell SK, Shatz CJ (1990) Functional synaptic circuits in the subplate during fetal and early postnatal development of cat visual cortex. J Neurosci 10:2601–2613

Friocourt G, Koulakoff A, Chafey P, Boucher D, Fauchereau F, Chelly J, Francis F (2003) Doublecortin functions at the extremities of growing neuronal processes. Cereb Cortex 13:620–626

Fukuchi-Shigomori T, Grove EA (2001) Neocortex patterning by the secreted signaling molecule FGF8. Science 294:1071–1074

Fukuchi-Shigomori T, Grove EA (2003) Emx2 patterns the neocortex by regulating FGF positional signaling. Nat Neurosci 6:825–831

Furuta Y, Piston DW Hogan BL (1997) Bone morphogenetic proteins (BMPs) as regulators of dorsal forebrain development. Development 124:2203–2212

Gabbott PLA, Somogyi P (1986) Quantitative distribution of GABA-immunoreactive neurons in the visual cortex (area 17) of the cat. Exp Brain Res 61:323–331

Gadisseux JF, Goffinet AM, Lyon G, Evrard P (1992) The human transient subpial granular layers: an optical, immunohistochemical, and ultrastructural analysis. J Comp Neurol 324:94–114

Gage FH (2000) Mammalian neural stem cells. Science 287:1433–1438

Gambello MJ, Darling DL, Yingling J, Tanaka T, Gleeson JG, Wynshaw-Boris A (2003) Multiple dose-dependent effects of Lis1 on cerebral cortical development. J Neurosci 23:1719–1729

Garel C, Chantrel E, Brisse H, Elmaleh M, Luton D, Oury JF, Sebag G, Hassan M (2001) Fetal cerebral cortex: normal gestational landmarks identified using prenatal MR imaging. AJNR Am J Neuroradiol 22:184–189

Garel S, Huffman KJ, Rubenstein JL (2003) Molecular regionalization of the neocortex is disrupted in Fgf8 hypomorphic mutants. Development 130:1903–1914

Ghosh A, Shatz CJ (1992) Involvement of subplate neurons in the formation of ocular dominance columns. Science 255:1441–1443

Gilles FH, Gómez IG (2005) Developmental neuropathology of the second half of gestation. Early Human Development 81:245–253

Gilmore EC, Ohshima T, Goffinet AM, Kulkarni AB, Herrup K (1998) Cyclin-dependent kinase 5-deficient mice demonstrate novel developmental arrest in cerebral cortex. J Neurosci. 18:6370–6377

Gleeson JG, Allen KM, Fox JW, Lamperti ED, Berkovic S, Scheffer I, Cooper EC, Dobyns WB, Minnerath SR, Ross ME, Walsh CA (1998) Doublecortin, a brain-specific gene mutated in human X-linked lissencephaly and double cortex syndrome, encodes a putative signaling protein. Cell 92:63–72

Gleeson JG, Lin PT, Flanagan LA, Walsh CA (1999) Doublecortin is a microtubule-associated protein and is expressed widely by migrating neurons. Neuron 23:257–271

Gleeson JG, Walsh CA (2000) Neuronal migration disorders: from genetic diseases to developmental mechanisms. Trends Neurosci 23:352–359

Gleeson JG, Luo RF, Grant PE, Guerrini R, Huttenlocher PR, Berg MJ, Ricci S, Cusmai R, Wheless JW, Berkovic S, Scheffer I, Dobyns WB, Walsh CA (2000) Genetic and neuroradiological heterogeneity of double cortex syndrome. Ann Neurol 47:265–269

Goffinet AM (1984) Events governing organization of postmigratory neurons: studies on brain development in normal and reeler mice. Brain Res Rev 7:261–296

Goffinet AM, Bar I, Bernier B, Trujillo C, Raynaud A, Meyer G (1999) Reelin expression during embryonic brain development in lacertilian lizards. J Comp Neurol 414:533–550

Gomez-Isla T, Price JL, McKeel DW Jr, Morris JC, Growdon JH, Hyman BT (1996) Profound loss of layer II entorhinal cortex neurons occurs in very mild Alzheimer's disease. J Neurosci. 16:4491–4500

Gonchar Y, Burkhalter A (1997) Three distinct families of GABAergic neurons in rat visual cortex. Cereb Cortex 7:347–358

Gorski JA, Talley T, Qiu M, Puelles L, Rubenstein JL, Jones KR (2002) Cortical excitatory neurons and glia, but not GABAergic neurons, are produced in the Emx1-expressing lineage. J Neurosci 22:6309–6314

Graham ME, Ruma-Haynes P, Capes-Davies AG, Dunn JM, Tan TC, Valova VA, Robinson PJ, Jeffrey PL (2004) Multisite phosphorylation of doublecortin by cyclin-dependent kinase 5. Biochem J 381:471–481

Graus-Porta D, Blaess S, Senften M, Littlewood-Evans A, Damsky C, Huang Z, Orban P, Klein R, Schittny JC, Muller U (2001) Beta1-class integrins regulate the development of laminae and folia in the cerebral and cerebellar cortex. Neuron 31:367–379

Grove EA, Fukuchi-Shimogori T (2003) Generating the cerebral cortical area map. Annu Rev Neurosci 26:355–380

Grove EA, Tole S, Limon J, Yip L, Ragsdale CW (1998) The hem of the embryonic cerebral cortex is defined by the expression of multiple Wnt genes and is compromised in Gli3-deficient mice. Development 125:2315–2325

Grove EA, Williams BP, Li DQ, Hajihosseini M, Friedrich A, Price J (1993) Multiple restricted lineages in the embryonic rat cerebral cortex. Development 117:553–561

Guerrini R, Filippi T (2004) Neuronal migration disorders, genetics, and epileptogenesis. J Child Neurol 20:287–299

Guillemot F, Molnar Z, Tarabykin V, Stoykova A (2006) Molecular mechanisms of cortical differentiation. Europ J Neurosci (in press)

Guidotti A, Auta J, Davis JM, Di-Giorgi-Gerevini V, Dwivedi Y, Grayson DR, Impagnatiello F, Pandey G, Pesold C, Sharma R, Uzunov D, Costa E (2000) Decrease in reelin and glutamic acid decarboxylase67 (GAD67) expression in schizophrenia and bipolar disorder: a postmortem brain study. Arch Gen Psychiatry 57:1061–1069

Gulisano M, Broccoli V, Pardini C, Boncinelli E (1996) Emx1 and Emx2 show different patterns of expression during proliferation and differentiation of the developing cerebral cortex in the mouse. Eur J Neurosci 8:1037–1050

Gupta A, Sanada K, Miyamoto DT, Rovelstad S, Nadarajah B, Pearlman AL, Brunstrom J, Tsai LH (2003) Layering defect in p35 deficiency is linked to improper neuronal-glial interaction in radal migration. Nat Neurosci 6:1284–1291

Hamasaki T, Leingartner A, Ringstedt T, O'Leary DD (2004) EMX2 regulates sizes and positioning of the primary sensory and motor areas in neocortex by direct specification of cortical progenitors. Neuron 43:359–372

Hammond V, Howell B, Godinho L, Tan SS (2001) disabled-1 functions cell autonomously during radial migration and cortical layering of pyramidal neurons. J Neurosci 21:8798–8808

Hannan AJ, Servotte S, Katsnelson A, Sisodiya S, Blakemore C, Squier M, Molnar Z (1999) Characterization of nodular neuronal heterotopia in children. Brain 122:219–238

Harding, B (1996) Gray matter heterotopia. In: Dysplasias of cerebral cortex and epilepsy. R. Guerrini, F. Andermann, R. Canapicchi, J. Roger, B. Zilfkin, and P. Pfanner, eds. (Philadelphia-New York: Lippincott-Raven), pp 81–88

Hartfuss E, Galli R, Heins N, Gotz M (2001) Characterization of CNS precursor subtypes and radial glia. Dev Biol 229:15–30

Hartfuss E, Forster E, Bock HH, Hack MA, Leprince P, Luque JM, Herz J, Frotscher M, Gotz M (2003) Reelin signaling directly affects radial glia morphology and biochemical maturation. Development 130:4597–4609

Hartmann H, Uyanik G, Gross C, Hehr U, Lucke T, Arslan-Kirchner M, Antosch B, Das AM, Winkler J (2004) Agenesis of the corpus callosum, abnormal genitalia and intractable epilepsy due to a novel familial mutation in the Aristaless-related homeobox gene. Neuropediatrics 35:157–160

Hatanaka Y, Murakami F (2002) In vitro analysis of the origin, migratory behavior, and maturation of cortical pyramidal cells. J Comp Neurol 454:1–14

Hattori M, Adachi H, Tsujimoto M, Arai H, Inoue K (1994) Miller-Dieker lissencephaly gene encodes a subunit of brain platelet-activating factor acetylhydrolase. Nature 370:216–218

Haubensak W, Attardo A, Denk W, Huttner WB (2004) Neurons arise in the basal neuroepithelium of the early mammalian telencephalon: A major site of neurogenesis. Natl Acad Sci USA 101:3196–3201

Herz J, Bock HH (2002) Lipoprotein receptors in the nervous system. Annu Rev Biochem 71:405–34

Hevner RF (2000) Development of connections in the human visual system during fetal mid-gestation: A DiI tracing study. J Neuropathol Exp Neurol 59:385–392

Hevner RF, Daza RAM, Englund C, Kohtz, Fink A (2004) Postnatal shifts of interneuron position in the neocortex of normal and reeler mice: Evidence for inward radial migration. Neuroscience 124:605–618

Hevner RF, Neogi T, Englund C, Daza RA, Fink A (2003) Cajal-Retzius cells in the mouse: transcription factors, neurotransmitters, and birthdays suggest a pallial origin. Brain Res Dev Brain Res 141:39–53

Hevner RF, Shi L, Justice N, Hsueh Y, Sheng M, Smiga S, Bulfone A, Goffinet AM, Campagnoni AT, Rubinstein JL (2001) Tbr1 regulates differentiation of the preplate and layer 6. Neuron 29:353–366

Hiesberger T, Trommsdorff M, Howell BW, Goffinet A, Mumby MC, Cooper JA and Herz J (1999) Direct binding of Reelin to VLDL receptor and ApoE receptor 2 induces tyrosine phosphorylation of disabled-1 and modulates tau phosphorylation. Neuron 24:481–489

Hill RS, Walsh CA (2005) Molecular insights into human brain evolution. Nature 437:64–67

Hirotsune S, Fleck MW, Gambello MJ, Bix GJ, Chen A, Clark GD, Ledbetter DH, McBain CJ, Wynshaw-Boris A (1998) Graded reduction of Pafah1b1 (Lis1) activity results in neuronal migration defects and early embryonic lethality. Nat Genet 19:333–339

Hof PR, Glezer II, Nimchinsky EA, Erwin JM (2000) Neurochemical and cellular specializations in the mammalian neocortex reflect phylogenetic relationships: evidence from primates, cetaceans and artiodactyls. Brain Behav Evol 55:300–310

Hong SE, Shugart YY, Huang DT, Shahwan SA, Grant PE, Hourihane JO, Martin ND, Walsh CA (2000) Autosomal recessive lissencephaly with cerebellar hypoplasia is associated with human RELN mutations. Nat Genet 26:93–96

Honig LS, Herrmann K, Shatz CJ (1996) Developmental changes revealed by inmunohisto-chemical markers in human cerebral cortex. Cereb Cortex 6:794–806

Horesh D, Sapir T, Francis F, Wolf SG, Caspi M, Elbaum M, Chelly J, Reiner O (1999) Doublecortin, a stabilizer of microtubules. Hum Mol Genet 8:1599–1610

Howard A, Tamas G, Soltesz I (2005) Lighting the chandelier: new vistas for axo-axonic cells. Trends Neurosci 28:310–316

Howell BW, Hawkes R, Soriano P, Cooper JA (1997) Neuronal position in the developing brain is regulated by mouse disabled-1. Nature 389:733–737

Howell BW, Herrick TM, Cooper JA (1999) Reelin-induced tyrosine phosphorylation of disabled 1 during neuronal positioning. Genes Dev 13:643–648

Howell BW, Lanier LM, Frank R, Gertler FB, Cooper JA (1999) The disabled 1 phosphotyrosine-binding domain binds to the internalization signals of transmem-brane glycoproteins and to phospholipids. Mol Cell Biol 19:5179–5188

Howell BW, Herrick TM, Hildebrand JD, Zhang Y, Cooper JA (2000) Dab1 tyrosine phos-phorylation sites relay positional signals during mouse brain development. Curr Biol 10:877–885

Huffman KJ, Garel S, Rubenstein JL (2004) Fgf8 regulates the development of intra-neocortical projections. J Neurosci 24:8917–8923

Hunter-Schaedle KE (1997) Radial glial cell development and transformation are disturbed in reeler forebrain. J Neurobiol 33:459–472

Huntley GW, Jones EG (1990) Cajal-Retzius in the developing monkey neocortex show immunoreactivity for calcium-binding proteins. J Neurocytol 19:200–212

Ignatova N, Sindic CJ, Goffinet AM (2004) Characterization of the various forms of the Reelin protein in the cerebrospinal fluid of normal subjects and in neurological diseases. Neurobiol Dis 15:326–330

Ikeda Y, Terashima T (1997) Expression of reelin, the gene responsible for the reeler muta-tion, in embryonic development and adulthood in the mouse. Dev Dyn 210:157–172

Impagnatiello F, Guidotti AR, Pesold C, Dwivedi Y, Caruncho H, Pisu MG, Uzunov DP, Smalheiser NR, Davis JM, Pandey GN, Pappas GD, Tueting P, Sharma RP, Costa E (1998) A decrease of reelin expression as a putative vulnerability factor in schizophrenia. Proc Natl Acad Sci U S A 95:15718–15723

Irwin M, Marin MC, Phillips AC, Seelan RS, Smith DI, Liu W, Flores ER, Tsai KY, Jacks T, Vousden KH, Kaelin WG Jr (2000) Role for the p53 homologue p73 in E2F-1-induced apoptosis. Nature 407:645–648

Jackson AP, Eastwood H, Bell SM, Adu J, Toomes C, Carr IM, Roberts E, Hampshire DJ, Crow YJ, Mighell AJ, Karbani G, Jafri H, Rashid Y, Mueller RF, Markham AF, Woods CG (2002) Identification of microcephalin, a protein implicated in determining the size of the human brain. Am J Hum Genet 71:136–142

Jackson CA, Peduzzi JD, Hickey TL (1989) Visual development in the ferret. I. Genesis and migration of visual cortical neurons. J Neurosci 9:1242–1253

Jellinger K, Rett A (1976) Agyria-pachygyria (lissencephaly syndrome). Neuropädiatrie 7:66–91

Jessberger S, Romer B, Babu H, Kempermann G (2005) Seizures induce proliferation and dispersion of doublecortin-positive hippocampal progenitor cells. Exp Neurol 196:342–351

Jossin Y (2004) Neuronal migration and the role of Reelin during early development of the cerebral cortex. Mol Neurobiol 30:225–251

Jossin Y, Bar I, Ignatova N, Tissir F, De Rouvroit CL, Goffinet AM (2003) The reelin signaling pathway: some recent developments. Cereb Cortex 13:627–633

Jossin Y, Goffinet AM (2001) Reelin does not directly influence axonal growth. J Neurosci 21:1–4

Jossin Y, Ignatova N, Hiesberger T, Herz J, Lambert de Rouvroit C, Goffinet AM (2004) The central fragment of Reelin, generated by proteolytic processing in vivo, is critical to its function during cortical plate development. J Neurosci 24:514–521

Jossin Y, Ogawa M, Metin C, Tissir F, Goffinet AM (2003) Inhibition of SRC family kinases and non-classical protein kinases C induce a reeler-like malformation of cortical plate development. J Neurosci 23:9953–9959

Jost CA, Marin MC, Kaelin WG (1997) p73 is a simian (correction of human) p53 related protein that can induce apoptosis. Nature 389:191–194

Kaghad M, Bonnet H, Yang A, Creancier L, Biscan JC, Valent A, Minty A, Chalon P, Lelias JM, Dumont X, Ferrara P, McKeon F, Caput D (1997) Monoallelically expressed gene related to p53 at 1p36, a region frequently deleted in neuroblastoma and other human cancers. Cell 90:809–819

Kammermeier L, Reichert H (2001) Common developmental genetic mechanisms for patterning invertebrate and vertebrate brains. Brain Res Bull 55:675–682

Kanold PO, Kara P, Reid RC, Shatz CJ (2003) Role of subplate neurons in functional maturation of visual cortical columns. Science 301:521–525

Kato M, Das S, Petras K, Kitamura K, Morohashi K, Abuelo DN, Barr M, Bonneau D, Brady AF, Carpenter NJ, Cipero KL, Frisone F, Fukuda T, Guerrini R, Iida E, Itoh M, Lewanda AF, Nanba Y, Oka A, Proud VK, Saugier-Veber P, Schelley SL, Selicorni A, Shaner R, Silengo M, Stewart F, Sugiyama N, Toyama J, Toutain A, Vargas AL, Yanazawa M, Zackai EH, Dobyns WB (2004) Mutations of ARX are associated with striking pleiotropy and consistent genotype-phenotype correlation. Hum Mutat 23:147–159

Kato M, Dobyns WB (2005) X-linked lissencephaly with abnormal genitalia as a tangential migration disorder causing intractable epilepsy: proposal for a new term, "interneuronopathy". J Child Neurol 20:392–397

Kennedy H, Dehay C (1993) Cortical specification of mice and men. Cereb Cortex 3:171–186

Keshvara L, Benhayon D, Magdaleno S, Curran T (2001) Identification of reelin-induced sites of tyrosine phosphorylation on disabled 1. J Biol Chem 276:16008–16014

Keshvara L, Magdaleno S, Benhayon D, Curran T (2002) Cyclin-dependent kinase 5 phosphorylates disabled 1 independently of Reelin signaling. J Neurosci 22:4869–4877

Kitamura K, Yanazawa M, Sugiyama N, Miura H, Iizuka-Kogo A, Kusaka M, Omichi K, Suzuki R, Kato-Fukui Y, Kamiirisa K, Matsuo M, Kamijo S, Kasahara M, Yoshioka H, Ogata T, Fukuda T, Kondo I, Kato M, Dobyns WB, Yokoyama M, Morohashi K (2002) Mutation of ARX causes abnormal development of forebrain and testes in mice and X-linked lissencephaly with abnormal genitalia in humans. Nat Genet 32:359–369

Koizumi H, Tanaka T, Gleeson JG (2006) Doublecortin-like kinase functions with doublecortin to mediate fiber tract decussation and neuronal migration. Neuron 49:55–66

Kolk SM, Whitman MC, Yun ME, Shete P, Donoghue MJ (2005) A unique subpopulation of Tbr1-expressing deep layer neurons in the developing cerebral cortex. Mol Cell Neurosci 30:538–551

König N, Roch G, Marty R (1975) The onset of synaptogenesis in rat temporal cortex. Anat Embryol 148:73–87

König N, Valat J, Fulcrand J, Marty R (1977) The time of origin of Cajal-Retzius cells in the rat temporal cortex An autoradiographic study. Neurosci Lett 4:21–26

Kornack DR, Rakic P (1998) Changes in cell-cycle kinetics during the development and evolution of primate neocortex. Proc Natl Acad Sci USA 95:1242–1246

Kostovic I (1986) Prenatal Development of nucleus basalis complex and related fibre system in man: a histochemical study. Neurosci 17:1047–1077

Kostovic I (1990a) Zentralnervensystem. In: Humanembryologie. Ed KV Hinrichsen, Springer Verlag Berlin, pp 381–446

Kostovic I (1990b) Structural and histochemical reorganization of the human prefrontal cortex during perinatal and postnatal life. Prog Brain Res 85:223–239

Kostovic I, Goldman-Rakic P (1983) Transient cholinesterase staining in the mediodorsal nucleus of the thalamus and its connections in the developing human and monkey brain. J Comp Neurol 219:431–447

Kostovic I, Judas M, Rados M, Hrabac P (2002) Laminar organization of the human fetal cerebrum revealed by histochemical markers and magnetic resonance imaging. Cereb cortex 12:536–544

Kostovic I, Molliver ME (1974) A new interpretation of the laminar development of the cerebral cortex: synaptogenesis in different layers of neopallium in the human fetus. Anat Rec 178:395

Kostovic I, Rakic P (1984) Development of prestriate visual projections in the monkey and human fetal cerebrum revealed by transient cholinesterase staining. J Neurosci 4:25–42

Kostovic I, Rakic P (1980) Cytology and time of origin of interstitial neurons in the white matter in infant and adult human and monkey telencephalon. J Neurocytol 9:219–242

Kostovic I, Rakic P (1990) Developmental history of the transient subplate zone in the visual and somatosensory cortex of the macaque monkey and human brain. J Comp Neurol 297:441–470

Kriegstein AR, Noctor SC (2004) Patterns of neuronal migration in the embryonic cortex. Trends Neurosci 27:392–399

Krmpotic-Nemanic J, Kostovic I, Vidic Z, Nemanic D, Kostovic-Knezevic L (1987) Development of Cajal-Retzius cells in the human auditory cortex. 103:477–480

Kuo G, Arnaud L, Kronstad-O'Brian P, Cooper JA (2005) Absence of Fyn and Src causes a reeler-like phenotype. J Neurosci 25:8578–8586

Kubota Y, Hattori R, Yui Y (1994) Three distinct subpopulations of GABAergic neurons in rat frontal agranular cortex. Brain Res 649:159–173

Kwon YT, Tsai LH (1998) A novel disruption of cortical development in p35(–/–) mice distinct from reeler. J Comp Neurol 395:510–522

Lacor PN, Grayson DR, Auta J, Suguya I, Costa E, Guidotti A (2000) Reelin secretion from glutamatergic neurons in culture is independent from neurotransmitter regulation. Proc Natl Acad Sci USA 97:3556–3561

Lambert de Rouvroit C, Goffinet AM (1998a) The reeler mouse as a model of brain development. Adv Anat Embryol Cell Biol 150:1–108

Lambert de Rouvroit C, Goffinet AM (1998b) Cloning of human DAB1and mapping to chromosome 1p31-p32. Genomics 53:246–247

Lambert de Rouvroit C, de Bergeyck V, Cortvrindt C, Bar I, Eeckhout Y, Goffinet AM (1999) Reelin, the extracellular matrix protein deficient in reeler mutant mice, is processed by a metalloproteinase. Exp Neurol 156:214–217

Lambert de Rouvroit C, Bernier B, Royaux I, de Bergeyck V, Goffinet AM (1999) Evolutionarily conserved, alternative splicing of reelin during brain development. Exp Neurol 156:229–238

Lavdas AA, Grigoriou M, Pachnis V, Parnavelas JG (1999) The medial ganglionic eminence gives rise to a population of early neurons in the developing cerebral cortex. J Neurosci 19:7881–7888

Ledbetter SA, Kuwano A, Dobyns WB, Ledbetter DH (1992) Microdeletions of chromosome 17p13 as a cause of isolated lissencephaly. Am J Hum Genet 50:182–189

Lee SM, Tole S, Grove E, McMahon AP (2000) A local Wnt-3a signal is required for development of the mammalian hippocampus. Development 127:457–467

Letinic K, Zoncu R, Rakic P (2002) Origin of GABAergic neurons in the human neocortex. Nature 417:645–649

Levine D, Barnes PD (1999) Cortical maturation in normal and abnormal fetuses as assessed with prenatal MR imaging. Radiology 210:751–758

Li BS, Sun MK, Zhang L, Takahashi S, Ma W, Vinade L, Kulkarni AB, Brady RO, Pant HC (2001) Regulation of NMDA receptors by cyclin-dependent kinase-5. Proc Natl Acad Sci U S A. 98:12742–12747

Lin PT, Gleeson JG, Corbo JC, Flanagan L, Walsh CA (2000) DCAMKL1 encodes a protein kinase with homology to doublecortin that regulates microtubule polymerization. J Neurosci 20:9152–9161

Lindsay S, Sarma S, Martinez-de-la-Torre M, Kerwin J, Scott M, Luis Ferran J, Baldock R, Puelles L (2005) Anatomical and gene expression mapping of the ventral pallium in a three-dimensional model of developing human brain. Neuroscience 136:625–632

Liu WS, Pesold C, Rodriguez MA, Carboni G, Auta J, Lacor P, Larson J, Condie BG, Guidotti A, Costa E (2001) Down-regulation of dendritic spine and glutamic acid decarboxylase 67 expressions in the reelin haploinsufficient heterozygous reeler mouse. Proc Natl Acad Sci USA 98:3477–3482

Lo Nigro C, Chong CS, Smith AC, Dobyns WB, Carrozo R, Ledbetter DH (1997) Point mutations and an intragenic deletion in LIS1, the lissencephaly causative gene in isolated lissencephaly sequence and Miller-Dieker syndrome. Hum Mol Genet 6:157–164

Lo Turco JJ, Kriegstein AR (1991) Clusters of coupled neuroblasts in embryonic neocortex. Science 252:563–566

Lowell CA, Soriano P (1996) Knockouts of Src-family kinases: stiff bones, wimpy T cells, and bad memories. Genes Dev 10:1845–1857

Lukaszewicz A, Savatier P, Cortay V, Giroud P, Huissoud C, Berland M, Kennedy H, Dehay C (2005) G1 phase regulation, area-specific cell cycle control, and cytoarchitectonics in the primate cortex. Neuron 47:353–364

Luque JM, Morante-Oria J, Fairen A (2003) Localization of ApoER2, VLDLR and Dab1 in radial glia: groundwork for a new model of reelin action during cortical development. Brain Res Dev Brain Res 140:195–203

Luskin MB, Pearlman AL, Sanes JR (1988) Cell lineage in the cerebral cortex of the mouse studied in vivo and in vitro with a recombinant retrovirus. Neuron 1:635–647

Luskin MB, Shatz CJ (1985) Studies of the earliest generated cells of the cat's visual cortex: Cogeneration of the cells of the marginal zone and subplate. J Neurosci 5:1062–1075

Luo L (2000) Rho GTPases in neuronal morphogenesis. Nature Rev Neurosci 1:173–180

McManus MF, Golden JA (2005) Neuronal migration in developmental disorders. J Child Neuro 20:280–286

McManus MF, Nasrallah IM, Pancoast MM, Wynshaw-Boris A, Golden JA (2004). Lis1 is necessary for normal non-radial migration of inhibitory interneurons. Am J Pathol. 165:775–784

Magdaleno S, Keshvara L, Curran T (2002) Rescue of ataxia and preplate splitting by ectopic expression of Reelin in *reeler* mice. Neuron 33:573–586

Mailhos C, Andre S, Mollereau B, Goriely A, Hemmati-Brivanlou A, Desplan C (1998) Related Drosophila Goosecoid requires a conserved heptapeptide for repression of paired-class homeoprotein activators. Development 125:937–947

Malatesta P, Hartfuss E, Gotz M (2000) Isolation of radial glial cells by fluorescent-activated cell sorting reveals a neuronal lineage. Development 127:5253–5263

Mallamaci A, Iannone R, Briata P, Pintonello L, Mercurio S, Boncinelli E, Corte G (1998) EMX2 protein in the developing mouse brain and olfactory area. Mech Dev 77:165–172

Mallamaci A, Muzio L, Chan CH, Parnavelas J, Boncinelli E (2000) Area identity shifts in the early cerebral cortex of Emx2–/– mutant mice. Nat Neurosci 3:679–686

Mallamaci A, Stoykova A. (2006) Gene networks controlling early cerebral cortex arealization. Eur. J. Neurosci. (In Press)

Marin O, Yaron A, Bagri A, Tessier-Lavigne M, Rubenstein JL (2001) Sorting of striatal and cortical interneurons regulated by semaphorin-neuropilin interactions. Science 293:872–875

Marín O, Rubenstein JL (2001) A long, remarkable journey: tangential migration in the telencephalon. Nat Rev Neurosci 2:780–790

Marín-Padilla M (1970) Prenatal and early postnatal ontogenesis of the human motor cortex: a Golgi study I The sequential development of the cortical layers. Brain Research 23:167–183

Marín-Padilla M (1971) Early prenatal ontogenesis of the cerebral cortex (neocortex) of the cat (Felis domestica). A Golgi study. I. The primordial neocortical organization. Z Anat Entwickl-Gesch 134:117–145

Marín-Padilla M (1972) Prenatal ontogenetic history of the principal neurons of the neocortex of the cat (Felis domestica). A Golgi study. II. Developmental differences and their significances. Z Anat Entwicklungsgesch 136:125–142

Marín-Padilla M (1978) Dual origin of the mammalian neocortex and evolution of the cortical plate. Anat Embryol 152:109–126

Marín-Padilla M (1990) Three-dimensional structural organization of layer 1 of the human cerebral cortex: A Golgi study. J Comp Neurol 299:89–105

Marin-Padilla M (1995) Prenatal development of fibrous (white matter), protoplasmic (gray matter), and layer I astrocytes in the human cerebral cortex: A Golgi study. J Comp Neurol 357:554–572

Marin-Padilla M, Marin-Padilla TM (1982) Origin, prenatal development and structural organization of layer I of the human cerebral (motor) cortex. A Golgi study. Anat Embryol (Berl) 164:161–206

Martin R, Gutierrez A, Penafiel A, Marin-Padilla M, de la Calle A (1999) Persistence of Cajal-Retzius cells in the adult human cerebral cortex. An immunohistochemical study. Histol Histopathol 14:487–890

Martinez-Cerdeno V, Clasca F (2002) Reelin immunoreactivity in the adult neocortex: a comparative study in rodents, carnivores, and non-human primates. Brain Res Bull 57:485–488

Martinez-Cerdeno V, Galazo MJ, Clasca F (2003) Reelin-immunoreactive neurons, axons, and neuropil in the adult ferret brain: evidence for axonal secretion of reelin in long axonal pathways. J Comp Neurol 463:92–116

Martinez-Cerdeno V, Galazo MJ, Cavada C, Clasca F (2002) Reelin immunoreactivity in the adult primate brain: intracellular localization in projecting and local circuit neurons of the cerebral cortex, hippocampus and subcortical regions. Cereb Cortex 12:1298–1311

McConnell SK, Ghosh A, Shatz CJ (1989) Subplate neurons pioneer the first axon pathways from the cerebral cortex. Science 245:978–982

McManus MF, Nasrallah IM, Pancoast MM, Wynshaw-Boris A, Golden JA (2004) Lis1 is necessary for normal non-radial migration of inhibitory interneurons. Am J Pathol 165:775–784

Menezes JR, Smith CM, Nelson KC, Luskin MB (1995) The division of neuronal progenitor cells during migration in the neonatal mammalian forebrain. Mol Cell Neurosci 6:496–508

Merkle FT, Tramontin AD, Garcia-Verdugo JM, Alvarez-Buylla A (2004) Radial glia give rise to adult neural stem cells in the subventricular zone. Proc Natl Acad Sci U S A 101:17528–17532

Métin C, Baudoin J-P, Rakić S, Parnavelas JG (2006) Cell and molecular mechanisms involved in the migration of cortical interneurons. Eur J Neurosci 23:894–900

Métin C, Denizot JP, Ropert N (2000) Intermediate zone cells express calcium-permeable AMPA receptors and establish close contact with growing axons. J Neurosci 20:696–708

Meyer G (1983) Axonal patterns and topography of short-axon neurons in visual areas 17, 18 and 19 of the cat. J Comp Neurol 220:405–438

Meyer G (1987) Form and spatial arrangement of neurons in the primary motor cortex of man. J Comp Neurol 262:402–428

Meyer G (2001) Human neocortical development: the importance of embryonic and early fetal events. Neuroscientist 7:303–314

Meyer G, Cabrera-Socorro A, Pérez-García CG, Martínez-Millán L, Walker N, Caput D (2004) Developmental roles of p73 in Cajal-Retzius cells and cortical patterning. J Neurosci 24:9878–9887

Meyer G, Castro R, Soria JM, Fairén A (1999) The subpial granular layer in the developing cerebral cortex of rodents. In: Mouse brain development (Goffinet AM, Rakic P, eds) Berlin: Springer-Verlag, 277–291

Meyer G, De Rouvroit CL, Goffinet AM, Wahle P (2003) Disabled-1 mRNA and protein expression in developing human cortex. Eur J Neurosci 17:517–525

Meyer G, Ferres Torres R (1984) Postnatal maturation of nonpyramidal cells in the visual cortex of the cat. J Comp Neurol 228:226–244

Meyer G, Goffinet AM (1998) Prenatal development of reelin-immunoreactive neurons in the human neocortex. J Comp Neurol 397:29–40

Meyer G, Goffinet AM, Fairen A (1999) What is a Cajal-Retzius cell? A reassessment of a classical cell type based on recent observations in the developing neocortex. Cereb Cortex 9:765–775

Meyer G, González-Hernández T (1993) Developmental changes in layer I of the human neocortex during prenatal life: a DiI-tracing and AChE and NADPH-d histochemistry study. J Comp Neurol 338:317–336

Meyer G, Gonzalez-Hernandez T, Galindo-Mireles D, Castañeyra-Perdomo A, Ferres-Torres R (1991) The efferent projections of neurons in the white matter of different cortical areas of the adult rat. Anat Embryol 184:99–102

Meyer G, Pérez-García, CG, Abraham, H, Caput, D (2002a) Expression of p73 and Reelin in the developing human cortex. J Neurosc 22:4973–4986

Meyer G, Pérez-García, CG, Gleeson, JG (2002b) Selective expression of Doublecortin and Lis1 in the developing human cortex suggests unique modes of neuronal movement. Cerebral Cortex 12:1225–1236

Meyer G, Schaaps JP, Moreau L, Goffinet AM (2000) Embryonic and early fetal development of the human neocortex. J Neurosci 20:1858–1868

Meyer G, Soria JM, Martinez-Galan JR, Martin-Clemente B, Fairen A (1998) Different origins and developmental histories of transient neurons in the marginal zone of the fetal and neonatal rat cortex. J Comp Neurol 397:493–518

Meyer G, Wahle P (1988) The early postnatal development of CCK-ir structures in the visual cortex of the cat. J Comp Neurol 276:360–386

Meyer G, Wahle P (1999) The paleocortical ventricle is the origin of reelin-expressing neurons in the marginal zone of the foetal human neocortex. Eur J Neurosci 11:3937–3944

Meyer G, Wahle P, Castaneyra-Perdomo A, Ferres-Torres R (1992) Morphology of neurons in the white matter of the adult human neocortex. Exp Brain Res 88:204–212

Miller MW (1986) The migration and neurochemical differentiation of gamma-amino-butyric acid (GABA)-immunoreactive neurons in rat visual cortex as demonstrated by a combined immunocytochemical-autoradiographic technique. Brain Res 393:41–46

Miyashita-Lin EM, Hevner R, Wassarman KM, Martinez S, Rubenstein JL (1999) Early neocortical regionalization in the absence of thalamic innervation. Science 285:906–909

Miyata T, Kawaguchi A, Okano H, Ogawa M (2001) Asymmetric inheritance of radial glial fibers by cortical neurons. Neuron 31:727–741

Mizuguchi M, Takashima S, Kakita A, Yamada M, Ikeda K (1995) Lissencephaly gene product. Localization in the central nervous system and loss of immunoreactivity in Miller-Dieker syndrome. Am J Pathol 147:1142–1151

Mollgard K, Balslev Y, Lauritzen B, Saunders NR (1987) Cell junctions and membrane specializations in the ventricular zone (germinal matrix) of the developing sheep brain: a CSF-brain barrier. J Neurocytol 16:433–444

Molliver ME, Kostovic I, Van der Loos H (1973) The development of synapses in cerebral cortex of the human fetus. Brain Res 50:403–407

Molnar M, Potkin SG, Bunney WE, Jones EG (2003) mRNA expression patterns and distribution of white matter neurons in dorsolateral prefrontal cortex of depressed patients differ from those in schizophrenia patients. Biol Psychiatry 53:39–47

Molnár Z, Adams R, Goffinet AM, Blakemore C (1998) The role of the first postmitotic cortical cells in the development of thalamocortical innervation in the *reeler* mouse. J Neurosci 18:5746–5765

Molnár Z, Métin C, Stoykova A, Tarabykin V, Price DJ, Francis F, Meyer G, Dehay C, Kennedy H (2006) Comparative aspects of cerebral cortical development. Eur. J. Neurosci. 23:921–934

Monuki ES, Porter FD, Walsh CA (2001) Patterning of the dorsal telencephalon and cerebral cortex by a roof plate – Lhx2 pathway. Neuron 32:591–604

Monuki ES, Walsh CA (2001) Mechanisms of cerebral cortical patterning in mice and humans. Nat Neurosci 4:1199–1206

Morante-Oria J, Carleton A, Ortino B, Kremer EJ, Fairen A, Lledo P-M (2003) Subpallial origin of a population of projecting pioneer neurons during corticogenesis. Natl Acad Sci USA 100:12468–12473

Morris NR (2000) Nuclear migration: From fungi to the mammalian brain. J Cell Biol 148:1097–1101

Morris NR, Efimov VP, Xiang X (1998) Nuclear migration, nucleokinesis and lissencephaly. Trends Cell Biol 8:467–470

Müller F, O'Rahilly R (1990) The human brain at stages 18–20, including the choroid plexuses and the amygdaloid and septal nuclei. Anat Embryol 182:285–306

Mrzljak L, Uylings HBM, Kostovic I, Van Eden CG (1988) Prenatal development of neurons in the human prefrontal cortex: I A qualitative Golgi study. J Comp Neurol 271:355–386

Mrzljak L, Uylings HBM, Van Eden CG, Judás M (1990) Neuronal development in human prefrontal cortex in prenatal and postnatal stages. Prog Brain Res 85:185–222

Muzio L, DiBenedetto B, Stoykova A, Boncinelli E, Gruss P, Mallamaci A (2002) Emx2 and Pax6 control regionalization of the pre-neuronogenic cortical primordium. Cereb Cortex 12:129–139

Muzio L, Mallamaci A (2005) Foxg1 confines Cajal-Retzius neurogenesis and hippocampal morphogenesis to the dorsomedial pallium. J Neurosci 25:4435–4441

Nadarajah B, Alifragis P, Wong ROL, Parnavelas JG (2002) Ventricle-directed migration in the developing cerebral cortex. Nature Neuroscience 5:218–224

Nadarajah B, Brunstrom JE, Grutzendler J, Wong RO, Pearlman AL (2001) Two modes of radial migration in early development of the cerebral cortex. Nat Neurosci 4:143–150

Nery S, Fishell G, Corbin JG (2002) The caudal ganglionic eminence is a source of distinct cortical and subcortical cell populations. Nat Neurosci 5:1279–1287

Niethammer M, Smith DS, Ayala R, Peng J, Ko J, Lee MS, Morabito M, Tsai LH (2000) NUDEL is a novel Cdk5 substrate that associates with LIS1 and cytoplasmic dynein. Neuron 28:697–711

Nikolic M, Chou MM, Lu W, Mayer BJ, Tsai LH (1998) The p35/Cdk5 kinase is a neuron-specific Rac effector that inhibits PAK 1 activity. Nature 395:194–198

Niu S, Renfro A, Quattrocchi CC, Sheldon M, D'Arcangelo G (2004) Reelin promotes hippocampal dendrite development through the VLDLR/ApoER2-Dab1 pathway. Neuron 41:71–84

Noctor SC, Flint AC, Weissman TA, Dammerman RS, Kriegstein AR (2001) Neurons derived from radial glial cells establish radial units in neocortex. Nature 409:714–720

Noctor SC, Flint AC, Weissman TA, Wong WS, Clinton BK, Kriegstein AR (2002) Dividing precursor cells of the embryonic cortical ventricular zone have morphological and molecular characteristics of radial glia. J Neurosci 22:3161–3173

Noctor SC, Martinez-Cerdeño V, Ivic L, Kriegstein AR (2004) Cortical neurons arise in symmetric and asymmetric division zones and migrate through specific phases. Nat Neurosci 7:136–144

Ogawa M, Miyata T, Nakajima K, Yagyu K, Seike M, Ikenaka K, Yamamoto H, Mikoshiba K (1995) The reeler gene-associated antigen on Cajal-Retzius neurons is a crucial molecule for laminar organization of cortical neurons. Neuron 14:899–912

Ohshima T, Ward JM, Huh CG, Longenecker G, Veeranna, Pant HC, Brady RO, Martin LJ, Kulkarni AB (1996) Targeted disruption of the cyclin-dependent kinase 5 gene results in abnormal corticogenesis, neuronal pathology and perinatal death. Proc Natl Acad Sci USA 93:11173–11178

Ohshima T, Ogawa M, Veeranna, Hirasawa M, Longenecker G, Ishiguro K, Pant HC, Brady RO, Kulkarni AB, Mikoshiba K (2001) Synergistic contributions of cyclin-dependant kinase 5/p35 and Reelin/Dab1 to the positioning of cortical neurons in the developing mouse brain. Proc Natl Acad Sci USA 98:2764–2769

Ohshima T, Mikoshiba K (2002) Reeling-signalling and Cdk5 in the control of neuronal positioning. Mol Neurobiol 26:153–166

Ohshima T, Ogura H, Tomizawa K, Hayashi K, Suzuki H, Saito T, Kamei H, Nishi A, Bibb JA, Hisanaga S, Matsui H, Mikoshiba K (2005) Impairment of hippocampal long-term depression and defective spatial learning and memory in p35 mice. J Neurochem 94:917–925

O'Leary DD, Nakagawa Y (2002) Patterning centers, regulatory genes and extrinsic mechanisms controlling arealization of the neocortex. Curr Opin Neurobiol 12:14–25

O'Rahilly R, Müller F (1994) The embryonic human brain. An atlas of developmental stages. Wiley-Liss, New York

O'Rahilly R, Müller F (2000) Prenatal ages and stages-measures and errors. Teratology 61:382–384

Palmini A, Andermann F, Olivier A, Tampieri D, Robitaille Y, Andermann E, Wright G (1991) Focal neuronal migration disorders and intractable partial epilepsy: a study of 30 patients. Ann Neurol 30:741–749

Pancoast M, Dobyns W, Golden JA (2005) Interneuron deficits in patients with the Miller-Dieker syndrome. Acta Neuropathol (Berl) 109:400–404

Pappas GD, Kriho V, Pesold C (2001) Reelin in the extracellular matrix and dendritic spines of the cortex and hippocampus: a comparison between wild type and heterozygous reeler mice by immunoelectron microscopy. J Neurocytol 30:413–425

Parnavelas JG (2000) The origin and migration of cortical neurones: new vistas. Trends Neurosci 23:126–131

Parnavelas JG, Nadarajah B (2001) Radial glial cells: are they really glia. Neuron 31:881–884

Peduzzi JD (1988) Genesis of GABA-immunoreactive neurons in the ferret visual cortex. J Neurosci 8:920–931

Pérez-Costas E, Melendez-Ferro M, Santos Y, Anadon R, Rodicio MC, Caruncho HJ (2002) Reelin immunoreactivity in the larval sea lamprey brain. J Chem Neuroanat 23:211–221

Pérez-García CG, González-Delgado FJ, Suárez-Solá ML, Castro-Fuentes R, Martín-Trujillo JM, Ferres-Torres R, Meyer G (2001) Reelin-immunoreactive neurons in the adult vertebrate pallium. J Chem Neuroanat 21:41–51

Pérez-García CG, Tissir F, Goffinet AM, Meyer G (2004) Reelin receptors in developing laminated brain structures of mouse and human. Eur J Neurosci 20:2827–2832

Pesold C, Impagnatiello F, Pisu MG, Uzunov DP, Costa E, Guidotti A, Caruncho HJ (1998) Reelin is preferentially expressed in neurons synthesizing gamma-aminobutyric acid in cortex and hippocampus of adult rats. Proc Natl Acad Sci USA 95:3221–3226

Pesold C, Liu WS, Guidotti A, Costa E, Caruncho HJ (1999) Cortical bitufted, horizontal, and Martinotti cells preferentially express and secrete reelin into perineuronal nets, nonsynaptically modulating gene expression. Proc Natl Acad Sci USA 96:3217–3222

Pilz DT, Matsumoto N, Minnerath S, Mills P, Gleeson JG, Allen KM, Walsh CA, Barkovich AJ, Dobyns WB, Ledbetter DH, Ross ME (1998) LIS1 and XLIS (DCX) mutations cause most classical lissencephaly, but different patterns of malformation. Hum Mol Genet 7:2029–2037

Pinard JM, Motte J, Chiron C, Brian R, Andermann E, Dulac O (1994) Subcortical laminar heterotopia and lissencephaly in two families: a single X linked dominant gene. J Neurol Neurosurg Psychiatry 57:914–920

Pinto-Lord MC, Evrard P, Caviness VS Jr (1982) Obstructed neuronal migration along radial glial fibers in the neocortex of the reeler mouse: a Golgi-EM analysis. Brain Res 256:379–393

Poirier K, Van Esch H, Friocourt G, Saillour Y, Bahi N, Backer S, Souil E, Castelnau-Ptakhine L, Beldjord C, Francis F, Bienvenu T, Chelly J (2004) Neuroanatomical distribution of ARX in brain and its localisation in GABAergic neurons. Brain Res Mol Brain Res 122:35–46

Polleux F, Whitford KL, Dijkhuizen PA, Vitalis T, Gosh A (2002) Control of cortical interneuron migration by neurotrophins and P13-kinase signaling. Development 129:3147–3160

Poljakow GI (1979) Entwicklung der Neuronen der menschlichen Gehirnrinde. VEB Georg Thieme Verlag Leipzig

Porteus MH, Bulfone A, Liu JK, Puelles L, Lo LC, Rubenstein JL (1994) DLX-2, MASH-1, and MAP-2 expression and bromodeoxyuridine incorporation define molecularly distinct cell populations in the embryonic mouse forebrain. J Neurosci 14:6370–6383

Pozniak CD, Radinovic S, Yang A, McKeon F, Kaplan DR, Miller FD (2000) An anti-apoptotic role for the p53 family member, p73, during developmental neuron death. Science 289:304–306

Pozniak CD, Barnabe-Heider F, Rymar VV, Lee AF, Sadikot AF, Miller FD (2002) p73 is required for survival and maintenance of CNS neurons. J Neurosci 22:9800–9809

Price M (1993) Members of the Dlx- and Nkx2-gene families are regionally expressed in the developing forebrain. J Neurobiol 24:1385–1399

Puelles L, Kuwana E, Puelles E, Bulfone A, Shimamura K, Keleher J, Smiga S, Rubenstein JL (2000) Pallial and subpallial derivatives in the embryonic chick and mouse telencephalon, traced by the expression of the genes Dlx-2, Emx-1, Nkx-2.1, Pax-6, and Tbr-1. J Comp Neurol 424:409–38

Puelles L, Rubenstein JL (1993) Expression patterns of homeobox and other putative regulatory genes in the embryonic mouse forebrain suggest a neuromeric organization. Trends Neurosci 16:472–479

Raedler E, Raedler A (1978) Autoradiographic study of early neurogenesis in rat neocortex. Anat Embryol 154:267–284

Ragsdale CW, Grove EA (2001) Patterning the mammalian cerebral cortex. Curr Opin Neurobiol 11:50–58

Rakic S, Zecevic N (2003) Emerging complexity in layer I of human cerebral cortex. Cereb Cortex 13:1072–1083

Rakic P (1972) Mode of cell migration to the superficial layers of fetal monkey neocortex. J Comp Neurol 145:61–83

Rakic P (1974) Neurons in rhesus monkey visual cortex: systematic relation between time of origin and eventual disposition. Science 183:425–427

Rakic P (1995) A small step for the cell, a giant leap for mankind: a hypothesis of neocortical expansion during evolution. Trend Neurosci 18:383–388

Rakic P (2003) Developmental and evolutionary adaptations of cortical radial glia. Cereb Cortex 13:541–549

Rakic P, Yakovlev PI (1968) Development of the corpus callosum and cavum septi in man. J Comp Neurol 132:45–72

Ramón y Cajal S (1891) Sur la structure de l'ecorce cérébrale de quelques mammifères. La Cellule 7:123–176

Ramón y Cajal S (1899a) Estudios sobre la corteza cerebral humana I Corteza Visual. Rev Trimestral Micrográfica 4:1–63

Ramón y Cajal S (1899b) Estudios sobre la corteza cerebral humana II Estructura de la corteza motriz del hombre y mamíferos superiores. Rev Trimestral micrográfica 4:117–200

Ramón y Cajal S (1911) Histologie du système nerveux de l'homme et des vertébrés. Maloine, Paris

Ramón y Cajal S (1917) Recuerdos de mi vida. Vol.2 Historia de mi labor cientifica. Moya Madrid, pp 345–346

Ramos-Moreno T, Galazo MJ, Porrero C, Martinez-Cerdeno V, Clasca F (2006) Extracellular matrix molecules and synaptic plasticity: immunomapping of intracellular and secreted Reelin in the adult rat brain Europ J Neurosci 23:401–22

Ranke O (1910) Beiträge zur Kenntnis der normalen und pathologischen Hirnrindenbildung. Beitr Pathol Anat Allg Pathol 47:51–125

Rash BG, Grove EA (2006) Area and layer patterning in the developing cerebral cortex. Curr Opin Neurobiol. (In press)

Reiner O, Carrozzo R, Shen Y, Wehnert M, Faustinella F, Dobyns WB, Caskey CT and Ledbetter DH (1993) Isolation of a Miller-Dieker lissencephaly gene containing G protein beta-subunit-like repeats. Nature 364:717–721

Retzius G (1893) Die Cajal'schen Zellen der Grosshirnrinde beim Menschen und bei Säugetieren. Biologische Untersuchungen, Neue Folge 5:1–8

Rice DS, Sheldon M, D'Arcangelo G, Nakajima K, Goldowitz D, Curran T (1998) Disabled-1 acts downstream of Reelin in a signaling pathway that controls laminar organization in the mammalian brain. Development 125:3719–3729

Rice DS, Curran T (2001) Role of the reelin signaling pathway in central nervous system development. Annu Rev Neurosci 24:1005–1039

Rickmann M, Chronwall M, Wolff JR (1977) On the development of non-pyramidal neurons and axons outside the cortical plate: The early marginal zone as a pallial anlage. Anat Embryol 151:285–307

Rickmann M, Wolff JR (1981) Differentiation of 'preplate' neurons in the pallium of the rat. Bibl Anat 19:142–146

Roberts RC, Xu L, Roche JK, Kirkpatrick B (2005) Ultrastructural localization of reelin in the cortex in post-mortem human brain. J Comp Neurol 482:294–308

Robertson RT, Annis CM, Baratta J, Haraldson S, Ingeman J, Kageyama GH, Kimm E, Yu J (2000) Do subplate neurons comprise a transient population of cells in developing neocortex of rats? J Comp Neurol 426:632–650

Rockel AJ, Hiorns RW, Powell TPS (1980) The basic uniformity in structure of the neocortex. Brain 103:221–244

Rodriguez MA, Caruncho HJ, Costa E, Pesold C, Liu WS, Guidotti A (2002) In Patas monkey, glutamic acid decarboxylase-67 and reelin mRNA coexpression varies in a manner dependent on layers and cortical areas. J Comp Neurol 451:279–288

Rodriguez MA, Pesold C, Liu WS, Kriho V, Guidotti A, Pappas GD, Costa E (2000) Colocalization of integrin receptors and reelin in dendritic spine postsynaptic densities of adult nonhuman primate cortex. Proc Natl Acad Sci U S A 97:3550–3555

Rogers JH (1992) Immunohistochemical markers in rat cortex: colocalization of calretinin and calbindin 28 kDa with neuropeptides and GABA. Brain Res 587:147–157

Ross ME, Walsh CA (2001) Human brain malformations and their lessons for neuronal migration. Annu Rev Neurosci 24:1041–1070

Royaux I, Lambert de Rouvirot C, D'Arcangelo G, Demirov D, Goffinet AM (1997) Genomic organization of the mouse *reelin* gene. Genomics 46:240–250

Samuelsen GB, Larsen KB, Bogdanovic N, Laursen H, Graen N, Larsen JF, Pakkenberg B (2003) The changing number of cells in the human fetal forebrain and its subdivisions: a stereological analysis. Cereb cortex 13:115–122

Sanada K, Gupta A, Tsai LH (2004) Disabled-1-regulated adhesion of migrating neurons to radial glial fiber contributes to neuronal positioning during early corticogenesis. Neuron 42:197–211

Sansom SN, Hebert JM, Thammongkol U, Smith J, Nisbet G, Surani MA, McConnell SK, Livesey FJ (2005) Genomic characterisation of a Fgf-regulated gradient-based neocortical protomap. Development 132:3947–3961

Sapir T, Elbaum M, Reiner O (1997) Reduction of microtubule catastrophe events by LIS1, platelet-activating factor acetylhydrolase subunit. EMBO J 16:6977–6984

Sapir T, Cahana A, Seger R, Nekhai S, Reiner O (1999) LIS1 is a microtubule-associated phosphoprotein. Eur J Biochem 265:181–188

Sapir T, Horesh D, Caspi M, Atlas R, Burgess HA, Wolf SG, Francis F, Chelly J, Elbaum M, Pietrokovski S, Reiner O (2000) Doublecortin mutations cluster in evolutionarily conserved functional domains. Hum Mol Genet 9:703–712

Sasaki S, Shionoya A, Ishida M, Gambello MJ, Yingling J, Wynshaw-Boris A, Hirotsune S (2000) A LIS1/NUDEL/cytoplasmic dynein heavy chain complex in the developing and adult nervous system. Neuron 28:681–696

Schaar BT, McConnel SK (2005) Cytoskeletal coordination during neuronal migration. PNAS 102:13652–13657

Schaar BT, Kinoshita K, McConnell SK (2004) Doublecortin microtubule affinity is regulated by a balance of kinase and phosphatase activity at the leading edge of migrating neurons. Neuron. 41:203–213

Schaffer K (1917) Uber normale und pathologische Hirnfurchung. Z f d g Neur u Psych 38:1–72

Schierle GS, Gander JC, D'Orlando C, Ceilo MR, Vogt Weisenhorn DM (1997) Calretinin-immunoreactivity during postnatal development of the rat isocortex: a qualitative and quantitative study. Cereb Cortex 7:130–142

Schiffmann SN, Bernier B, Goffinet AM (1997) Reelin mRNA expression during mouse brain development. Eur J Neurosci 9:1055–1071

Schmid RS, Anton ES (2003) Role of integrins in the development of the cerebral cortex. Cereb Cortex 13:219–224

Schmid RS, Jo R, Shelton S, Kreidberg JA, Anton ES (2005) Reelin, integrin and Dab1 interactions during embryonic cerebral cortical development. Cereb Cortex 15:1632–1636

Sheldon M, Rice DS, D'Arcangelo G, Yoneshima H, Nakajima K, Mikoshiba K, Howell BW, Cooper JA, Goldowitz D, Curran T (1997) Scrambler and yotari disrupt the disabled gene and produce a reeler-like phenotype in mice. Nature 389:730–733

Sheppard AM, Pearlman AL (1997) Abnormal reorganization of preplate neurons and their associated extracellular matrix: an early manifestation of altered neocortical development in the reeler mutant mouse. J Comp Neurol 378:173–179

Shimogori T, Banuchi V, Ng HY, Strauss JB, Grove EA (2004) Embryonic signaling centers expressing BMP, WNT and FGF proteins interact to pattern the cerebral cortex. Development 131:5639–5647

Shu T, Tseng HC, Sapir T, Stern P, Zhou Y, Sanada K, Fischer A, Coquelle FM, Reiner O, Tsai LH (2006) Doublecortin-like kinase controls neurogenesis by regulating mitotic spindles and M phase progression. Neuron 49:25–39

Sicca F, Kelemen A, Genton P, Das S, Mei D, Moro F, Dobyns WB, Guerrini R (2003) Mosaic mutations of the LIS1 gene cause subcortical band heterotopia. Neurology 61:1042–1046

Sidman RL, Rakic P (1973) Neuronal migration, with special reference to developing human brain: a review. Brain Res 62:1–35

Sidman RL, Rakic P (1982) Development of the human central nervous system. In: Histology and histopathology of the nervous system. Eds Haymaker W, Adams RD, Charles C, Thomas Publisher Springfield Illinois USA Vol 1:3–145

Smart IHM, Dehay C, Giroud P, Berland M, Kennedy H (2002) Unique morphological features of the proliferative zones and postmitotic compartments of the neural epithelium giving rise to striate and extrastriate cortex in the monkey. Cereb Cortex 12:37–53

Simeone A, Acampora D, Gulisano M, Stornaiuolo A, Boncinelli E (1992) Nested expression domains of four homeobox genes in developing rostral brain. Nature 358:687–690

Smalheiser NR, Costa E, Guidotti A, Impagnatiello F, Auta J, Lacor P, Kriho V, Pappas GD (2000) Expression of reelin in adult mammalian blood, liver, pituitary pars intermedia, and adrenal chromaffin cells. Proc Natl Acad Sci U S A 97:1281–1286

Smart IHM, Dehay C, Giroud P, Berland M, Kennedy H (2002) Unique morphological features of the proliferative zones and postmitotic compartments of the neuronal epithelium giving rise to striate and extrastriate cortex in the monkey. Cereb Cortex 12:37–53

Smith DS, Niethammer M, Ayala R, Zhou Y, Gambello MJ, Wynshaw-Boris A, Tsai LH (2000) Regulation of cytoplasmic dynein behaviour and microtubule organization by mammalian Lis1. Nat Cell Biol 2:767–775

Smith Fernández A, Pieau C, Repérant J, Boncinelli E, Wassef M (1998) Expression of the *Emx-1* and *Dlx-1* homeobox genes define three molecularly distinct domains in the telencephalon of mouse, chick, turtle and frog embryos: implications for the evolution of telencephalic subdivisions in amniotes. Development 125:2099–2111

Somogyi P, Freund TF, Cowey A (1982) The axo-axonic interneuron in the cerebral cortex of the rat, cat and monkey. Neuroscience 7:2577–2607

Soria JM, Fairen A (2000) Cellular mosaics in the rat marginal zone define an early neocortical territorialization. Cereb Cortex 10:400–412

Soriano P, Montgomery C, Geske R, Bradley A (1991) Targeted disruption of the c-src proto-oncogene leads to osteopetrosis in mice. Cell 64:693–702

Sossey-Alaoui K, Hartung AJ, Guerrini R, Manchester DK, Posar A, Puche-Mira A, Andermann E, Dobyns WB, Srivastava AK (1998) Human doublecortin (DCX) and the homologous gene in mouse encode a putative Ca^{2+}-dependent signaling protein which is mutated in human X-linked neuronal migration defects. Hum Mol Genet 7:1327–1332

Spreafico R, Arcelli P, Frassoni C, Canetti P, Giaccone G, Rizzuti T, Mastrangelo M, Bentivoglio M (1999) Development of layer 1 of the human cerebral cortex after midgestation: arquitectonic findings, immunocytochemical identification of neurons and glia, and in situ labeling of apoptotic cells. J Comp Neurol 410:126–142

Steindler DA, Cowell SA (1976) Reeler mutant mouse: maintenance of appropriate and reciprocal connections in the cerebral cortex and thalamus. Brain Res 105:386–393

Stoykova A, Fritsch R, Walther C, Gruss P (1996) Forebrain patterning defects in Small eye mutant mice. Development 122:3453–3465

Stoykova A, Treichel D, Hallonet M, Gruss P (2000) Pax6 modulates the dorsoventral patterning of the mammalian telencephalon. J Neurosci 20:8042–8050

Strasser V, Fasching D, Hauser C, Mayer H, Bock HH, Hiesberger T, Herz J, Weeber EJ, Sweatt JD, Pramatarova A, Howell B, Schneider WJ, Nimpf J (2004) Receptor clustering is involved in reelin signalling. Mol Cell Biol 24:1378–1386

Striedter GF (2005) Principles of brain evolution. Sinauer Associates, USA

Stromme P, Mangelsdorf ME, Shaw MA, Lower KM, Lewis SM, Bruyere H, Lutcherath V, Gedeon AK, Wallace RH, Scheffer IE, Turner G, Partington M, Frints SG, Fryns JP, Sutherland GR, Mulley JC, Gecz J (2002) Mutations in the human ortholog of Aristaless cause X-linked mental retardation and epilepsy. Nat Genet 30:441–445

Stühmer T, Puelles L, Ekker M, Rubenstein JL (2002) Expression from a Dlx gene enhancer marks adult mouse cortical GABAergic neurons. Cereb Cortex 12:75–85

Stumm RK, Zhou C, Ara T, Lazarini F, Dubois-Dalcq M, Nagasawa T, Hollt V, Schulz S (2003) CXCR4 regulates interneuron migration in the developing neocortex. J Neurosci 23:5123–5130

Sun T, Patoine C, Abu-Khalil A, Visvader J, Sum E, Cherry TJ, Orkin SH, Geschwind DH, Walsh CA (2005) Early asymmetry of gene transcription in embryonic human left and right cerebral cortex. Science 308:1794–1798

Super H, Del Rio JA, Martinez A, Perez-Sust P, Soriano E (2000) Disruption of neuronal migration and radial glia in the developing cerebral cortex following ablation of Cajal-Retzius cells. Cereb Cortex 10:602–613

Sussel L, Marin O, Kimura S, Rubenstein JL (1999) Loss of Nkx2.1 homeobox gene function results in a ventral to dorsal molecular respecification within the basal telencephalon: evidence for a transformation of the pallidum into the striatum. Development 126:3359–3370

Swan A, Nguyen T, Suter B (1999) Drosophila Lissencephaly-1 functions with Bic-D and dynein in oocyte determination and nuclear positioning. Nat Cell Biol 1:444–449

Szentagothai J (1978) The Ferrier Lecture: the neuron network of the cerebral cortex: a functional interpretation. Proc R Soc Lond B Biol Sci 201:219–248

Tabata H, Nakajima K (2003) Multipolar migration: the third mode of radial neuronal migration in the developing cerebral cortex. J Neurosci 23:9996–10001

Takahashi T, Goto T, Miyama S, Nowakowski RS, Caviness VS Jr (1999) Sequence of neuron origin and neocortical laminar fate: relation to cell cycle of origin in the developing murine cerebral wall. J Neurosci 19:10357–10371

Takahashi T, Nowakowski RS, Caviness VS Jr. (1995) The cell cycle of the pseudostratified epithelium of the embryonic murine cerebral wall. J. Neurosci 15:6046–6057

Takahashi T, Nowakowski RS, Caviness VS Jr (1993) Cell cycle parameters and patterns of nuclear movement in the neocortical proliferative zone of the fetal mouse. J Neurosci 13:820–833

Takiguchi-Hayashi K, Sekiguchi M, Ashigaki S, Takamatsu M, Hasegawa H, Suzuki-Migishima R, Yokoyama M, Nakanishi S, Tanabe Y (2004) Generation of reelin-positive marginal zone cells from the caudomedial wall of telencephalic vesicles. J Neurosci 24:2286–2295

Talamillo A, Quinn JC, Collinson JM, Caric D, Price DJ, West JD, Hill RE (2003) Pax6 regulates regional development and neuronal migration in the cerebral cortex. Dev Biol 255:151–163

Tamamaki N, Fujimori KE, Takauji R (1997) Origin and route of tangentially migrating neurons in the developing neocortical intermediate zone. J Neurosci 17:8313–8323

Tamamaki et al. (2003) Evidence that Sema3A and Sema3F regulate the migration of GBAergic neurons in the developing neocortex. J Comp Neurol 455:238–248

Tamamaki N, Nakamura K, Okamoto K, Kaneko T (2001) Radial glia is a progenitor of neocortical neurons in the developing cerebral cortex. Neurosci Res 41:51–60

Tanaka D, Nakaya Y, Yanagawa Y, Obata K, Murakami F (2003) Multimodal tangential migration of neocortical GABAergic neurons independent of GPI-anchored proteins. Development 130:5803–5813

Tanaka T, Serneo FF, Tseng HC, Kulkarni AB, Tsai LH, Gleeson JG (2004) Cdk5 phosphorylation of doublecortin ser 297 regulates its effect on neuronal migration. Neuron 41:215–227

Tanaka T, Serneo FF, Higgins C, Gambello MJ, Wynshaw-Boris A, Gleeson JG (2004) Lis1 and doublecortin function with dynein to mediate coupling of the nucleus to the centrosome in neuronal migration. J Cell Biol 165:709–721

Tang D, Wang JH (1996) Cyclin-dependent kinase 5 (Cdk5) and neuron-specific Cdk5 activators. Prog Cell Cycle Res 2:205–216

Tarabykin V, Stoykova A, Usman N, Gruss P (2001) Cortical upper layer neurons derive from the subventricular zone as indicated by Svet1 gene expression. Development 128:1983–1993

Taylor KR, Holzer AK, Bazan JF, Walsh CA, Gleeson JG (2000) Patient mutations in doublecortin define a repeated tubulin-binding domain. J Biol Chem 275:34442–34450

Terashima T, Inoue K, Inoue Y, Mikoshiba K, Tsukada Y (1983) Distribution and morphology of corticospinal tract neurons in reeler mouse cortex by the retrograde HRP method. J Comp Neurol 218:314–326

Terashima T, Inoue K, Inoue Y, Mikoshiba K, Tsukada Y (1985) Distribution and morphology of callosal commissural neurons within the motor cortex of normal and reeler mice. J Comp Neurol 232:83–98

Terashima T, Inoue K, Inoue Y, Mikoshiba K (1987) Thalamic connectivity of the primary motor cortex of normal and reeler mutant mice. J Comp Neurol 257:405–421

Theil T, Alvarez-Bolado G, Walter A, Ruther U (2002) Wnt and Bmp signalling cooperatively regulate graded Emx2 expression in the dorsal telencephalon. Development 129:3045–3054

Thom M, Martinian L, Parnavelas JG, Sisodiya SM (2004) Distribution of cortical interneurons in grey matter heterotopia in patients with epilepsy. Epilepsia 45:916–923

Tissir F, Lambert de Rouvroit C, Goffinet AM (2002) The role of reelin in the development and evolution of the cerebral cortex. Braz J Med Biol Res 35:1473–1484

Tissir F, Goffinet AM (2003) Reelin in brain development. Nat Rev Neurosci 4:496–505

Tissir F, Lambert De Rouvroit C, Sire JY, Meyer G, Goffinet AM (2003) Reelin expression during embryonic brain development in Crocodylus niloticus. J Comp Neurol 457:250–262

Tole S, Goudreau G, Assimacopoulos S, Grove EA (2000b) Emx2 is required for growth of the hippocampus but not for hippocampal field specification. J Neurosci 20:2618–2625

Tramontin AD, Garcia-Verdugo JM, Lim DA, Alvarez-Buylla A (2003) Postnatal development of radial glia and the ventricular zone (VZ): a continuum of the neural stem cell compartment. Cereb Cortex 13:580–587

Trimborn M, Bell SM, Felix C, Rashid Y, Jafri H, Griffiths PD, Neumann LM, Krebs A, Reis A, Sperling K, Neitzel H, Jackson AP (2004) Mutations in microcephalin cause aberrant regulation of chromosome condensation. Am J Hum Genet 75:261–266

Trommsdorff M, Gotthardt M, Hiesberger T, Shelton J, Stockinger W, Nimpf J, Hammer RE, Richardson JA, Herz J (1999) Reeler/Disabled-like disruption of neuronal migration in knockout mice lacking the VLDL receptor and ApoE receptor 2. Cell 97:689–701

Tsai LH, Takahashi T, Caviness VS Jr, Harlow E (1993) Activity and expression pattern of cyclin-dependent kinase 5 in the embryonic mouse nervous system. Development 119:1029–1040

Tsai LH, Delalle I, Caviness VS Jr, Chae T, Harlow E (1994) p35 is a neural-specific regulatory subunit of cyclin-dependent kinase 5. Nature 371:419–423

Tsai JW, Chen Y, Kriegstein AR, Vallee (2005) LIS1 RNA interference blocks neural stem cell division, morphogenesis, and motility at multiple stages. J Cell Biol 170:935–945

Tsukada M, Prokscha A, Oldekamp J, Eichele G. (2003) Identification of neurabin II as a novel doublecortin interacting protein. Mech Dev 120:1033–43

Tsukada M, Prokscha A, Ungewickell E, Eichele G. (2005) Doublecortin association with actin filaments is regulated by neurabin II. J Biol Chem. 280:11361–11368

Tueting P, Costa E, Dwivedi Y, Guidotti A, Impagnatiello F, Manev R, Pesold C (1999) The phenotypic characteristics of heterozygous reeler mouse. Neuroreport 10:1329–1334

Uddin M, Wildman DE, Liu G, Xu W, Johnson RM, Hof PR, Kapatos G, Grossman LI, Goodman M (2004) Sister grouping of chimpanzees and humans as revealed by genome-wide phylogenetic analysis of brain gene expressions profiles. Proc Natl Acad Sci USA 101:2957–2962

Ulfig N (2002) Calcium-binding proteins in the human developing brain. Adv Anat Embryol Cell Biol 165:1–95

Uyanik G, Aigner L, Martin P, Gross C, Neumann D, Marschner-Schafer H, Hehr U, Winkler J (2003) ARX mutations in X-linked lissencephaly with abnormal genitalia. Neurology 61:232–235

Utsunomiya-Tate N, Kubo K, Tate S, Kainosho M, Katayama E, Nakajima K, Mikoshiba K (2000) Reelin molecules assemble together to form a large protein complex, which is inhibited by the function-blocking CR-50 antibody. Proc Natl Acad Sci USA 97:9729–9734

Valverde F, López-Mascaraque L, Santacana M, De Carlos JA (1995) Persistence of early-generated neurons in the rodent subplate: assessment of cell death in neocortex during the early postnatal period. J Neurosci 15:5014–5024

Van Hoesen GW, Hyman BT (1990) Hippocampal formation: anatomy and the patterns of pathology in Alzheimer's disease. Prog Brain Res 83:445–457

Verney C, Derer P (1995) Cajal-Retzius neurons in human cerebral cortex at midgestation show immunoreactivity for neurofilament and calcium-binding proteins. J Comp Neurol 359:144–153

Viot G, Sonigo P, Simon I, Simon-Bouy B, Chadeyron F, Beldjord C, Tantau J, Martinovic J, Esculpavit C, Brunelle F, Munnich A, Vekemans M, Encha-Razavi F (2004) Neocortical neuronal arrangement in LIS1 and DCX lissencephaly may be different. Am J Med Genet A 126:123–128

Vogt BA (1991) The role of layer 1 in cortical function. In A Peters and EG Jones (eds): Cerebral Cortex, Vol 9, Normal and Altered States of Function. New York: Plenum Press 49–80

Von Economo C, Koskinas GN (1925) Die Cytoarchitektonik der Hirnrinde des erwachsenen Menschen. Julius Springer Wien and Berlin

Wahle P, Meyer G (1987) Morphology and quantitative changes of transient NPY-ir neuronal populations during early postnatal development of the cat visual cortex. J Comp Neurol 261:165–192

Wahle P, Meyer G (1989) Early postnatal development of vasoactive intestinal polypeptide- and peptide histidine isoleucine-immunoreactive structures in the cat visual cortex. J Comp Neurol 282:215–248

Wahle P, Meyer G, Wu JY, Albus K (1987) Morphology and axon terminal pattern of glutamate decarboxylase-immunoreactive cell types in the white matter of the cat occipital cortex during early postnatal development. Dev Brain Res 36:53–61

Walsh CA (1999) Genetic malformations of the human cerebral cortex. Neuron 23:19–29

Walsh CA, Goffinet AM (2000) Potential mechanisms of mutations that affect neuronal migration in man and mouse. Curr Opin Genet Dev 10:270–274

Walsh GS, Orike N, Kaplan DR, Miller FD (2004) The invulnerability of adult neurons: a critical role for p73. J Neurosci 24(43):9638–9647

Ware ML, Fox JW, Gonzalez JL, Davis NM, Lambert de Rouvroit C, Russo CJ, Chua SC Jr, Goffinet AM, Walsh CA (1997) Aberrant splicing of a mouse disabled homolog, mdab1, in the scrambler mouse. Neuron 19:239–249

Weeber EJ, Beffert U, Jones C, Christian JM, Forster E, Sweatt JD, Herz J (2002) Reelin and ApoE receptors cooperate to enhance hippocampal synaptic plasticity and learning. J Biol Chem 277:39944–39952

Weimer JM, Anton ES (2006) Doubling up on microtubule stabilizers: synergistic functions of doublecortin-like kinase and doublecortin in the developing cerebral cortex. Neuron 49:3–4

Weisenhorn DM, Prieto EW, Celio MR (1994) Localization of calretinin in cells of layer 1 (Cajal-Retzius cells) of the developing cortex of the rat. Dev Brain Res 82:293–297

Weissman T, Noctor SC, Clinton BK, Honig LS, Kriegstein AR (2003) Neurogenic radial glial cells in reptile, rodent and human: from mitosis to migration. J Neurosci 13:550–559

Wichterle, H, Alvarez-Dolado, M, Erskine, L, Alvarez-Buylla, A (2003) Permissive corridor and diffusible gradients direct medial ganglionic eminence cell migration to the neocortex. Proc. Natl. Acad. Sci. USA 100:727–732

Wichterle H, Garcia-Verdugo JM, Herrera DG, Alvarez-Buylla A (1999) Young neurons from medial ganglionic eminence disperse in adult and embryonic brain. Nat Neurosci 2:461–466

Wichterle H, Turnbull DH, Nery S, Fishell G, Alvarez-Buylla A (2001) In utero fate mapping reveals distinct migratory pathways and fates of neurons born in the mammalian basal forebrain. Development 128:3759–3771

Willnow TE, Nykjaer A, Herz J (1999) Lipoprotein receptors: new roles for ancient proteins. Nat Cell Biol 1:157–162

Witter MP, Groenewegen HJ (1984) Laminar origin and septotemporal distribution of entorhinal and perirhinal projections to the hippocampus in the cat. J Comp Neurol 224:371–385

Wittmann T, Waterman-Storer CM (2001) Cell motility: can Rho GTPases and microtubules point the way? J Cell Sci 114:3795–3803

Wood JG, Martin S, Price DJ (1992) Evidence that the earliest generated cells of the murine cerebral cortex form a transient population in the subplate and marginal zone. J Comp Neurol 290:118–140

Wynshaw-Boris A, Gambello MJ (2001) Lis1 and dynein motor function in neuronal migration and development. Genes Dev 15:639–651

Wu SX, Goebbels S, Nakamura K, Nakamura K, Kometani K, Minato N, Kaneko T, Nave KA, Tamamaki N (2005) Pyramidal neurons of upper cortical layers generated by NEX-positive progenitor cells in the subventricular zone. PNAS 102:17172–17177

Xiang X, Osmani AH, Osmani SA, Xin M, Morris NR (1995) NudF, a nuclear migration gene in Aspergillus nidulans, is similar to the human LIS1 gene required for neuronal migration. Mol Biol Cell 6:297–310

Xie Z, Sanada K, Samuels BA, Shih H, Tsai LH (2003) Serine 732 phosphorylation of FAK by Cdk5 is important for microtubule organization, nuclear movement and neuronal migration. Cell 114:469–482

Xu Q, Cobos I, De la Cruz E, Rubenstein JL, Anderson SA (2004) Origins of cortical interneuron subtypes. J Neurosci 24:2612–2622

Xu Q, Cobos I, De la Cruz E, Anderson SA (2003) Cortical interneuron fate determination: Diverse sources for distinct subtypes? Cereb Cortex 13:670–676

Yamazaki H, Sekiguchi M, Takamatsu M, Tanabe Y, Nakanishi S (2004) Distinct ontogenic and regional expressions of newly identified Cajal-Retzius cell-specific genes during neocorticogenesis. Proc Natl Acad Sci USA 101:14509–14514

Yang A, Walker N, Bronson R, Kaghad M, Oosterwegel M, Bonnin J, Vagner C, Bonnet H, Dikkes P, Sharpe A, McKeon F, Caput D (2000) p73-deficient mice have neurological, pheromonal and inflammatory defects but lack spontaneous tumours. Nature 404:99–103

Yang A, McKeon F (2000) p63 and p73: P53 mimics, menaces and more. Nat Rev Mol Cell Biol 1:199–207

Yabut O, Renfro A, Niu S, Swann JW, Marin O, D'Arcangelo G (2005) Abnormal laminar position and dendrite development of interneurons in the *reeler* forebrain. Mol Brain Res

Yau HJ, Wang HF, Lai C, Liu FC (2003) Neural development of the neuregulin receptor ErbB4 in the cerebral cortex and the hippocampus: preferential expression by interneurons tangentially migrating from the ganglionic eminences. Cereb Cortex 13:252–264

Yoneshima H, Nagata E, Matsumoto M, Yamada M, Nakajima K, Miyata T, Ogawa M, Mikoshiba K (1997) A novel neurological mutant mouse, yotari, which exhibits reeler-like phenotype but expresses CR-50 antigen/reelin. Neurosci Res 29:217–223

Yoshida M, Suda Y, Matsuo I, Miyamoto N, Takeda N, Kuratani S, Aizawa S (1997) Emx1 and Emx2 functions in development of dorsal telencephalon. Development 124:101–111

Yoshida M, Assimacopoulos S, Jones KR, Grove EA (2006) Massive loss of Cajal-Retzius cells does not disrupt neocortical layer order. Development 133:537–545

Yuasa S, Hattori K, Yagi T (2004) Defective neocortical development in Fyn-tyrosine-kinase-deficient mice. Neuroreport 15:819–822

Yun K, Potter S, Rubenstein JL (2001) Gsh2 and Pax6 play complementary roles in dorsoventral patterning of the mammalian telencephalon. Development 128:193–205

Yuste R, Bonhoeffer T (2001) Morphological changes in dendritic spines associated with long-term synaptic plasticity. Annu Rev Neurosci 24:1071–1089

Zecevic N (2004) Specific characteristic of radial glia in the human fetal telencephalon. Glia 48:27–35

Zecevic N, Chen Y, Filipovic R (2005) Contributions of cortical subventricular zone to the development of the human cerebral cortex. J Comp Neurol 491:109–122

Zecevic N, Miloscvic A (1997) Initial development of gamma-aminobutyric acid immunoreactivity in the human cerebral cortex. J Comp Neurol 380:495–506

Zhu Y, Li HS, Zhou L, Wu JY, Rao Y (1999) Cellular and molecular guidance of GABAergic neuronal migration from an extracortical origin to the neocortex. Neuron 23:473–485

Zimmer C, Tiveron MC, Bodmer R, Cremer H (2004) Dynamics of Cux2 expression suggests that an early pool of SVZ precursors is fated to become upper cortical layer neurons. Cereb Cortex 14:1408–1420

Subject Index

Printing: Krips bv, Meppel
Binding: Stürtz, Würzburg